Praise for

THE SECRET LIVES OF BATS

"*The Secret Lives of Bats* is a whirlwind adventure story and a top-shelf natural history page-turner. But perhaps most important, it tells the feel-good conservation success story of the century: how Merlin Tuttle changed the world's perception of bats from monsters to angels – by befriending people, then showing them the truth. Everyone who cares about animals must read this riveting book about a fearless, indestructible gentleman-adventurer and the beautiful, gentle bats he has studied, protected, and loved."
– Sy Montgomery, author of *The Soul of an Octopus* and *The Good Good Pig*

"*The Secret Life of Bats* is one of the best, most interesting books I've ever read. Tuttle has had scores of mind-bending, hair-raising adventures all over the world, and he's insightful as he shows us bats as sweet-tempered, intelligent, social, and amazingly capable. Our lives would be miserable without these fascinating creatures, yet most of us misunderstand them. This book will fix that. It's an honor to read a fascinating book by a brave and highly competent scientist about some of the most important creatures on earth."
– Elizabeth Marshall Thomas, author of *The Hidden Life of Dogs* and *The Tribe of Tiger*

"Grips and doesn't let go . . . bats lead long, complex, and wondrous lives under cover of darkness. I'm grateful we're given a peek at them in this uplifting and deeply satisfying book."
– *Wall Street Journal*

"The biggest secret in Merlin Tuttle's thrill-filled account of his bat-obsessed life is how he survived to tell his amazing tales. Tuttle is a legend in bat circles and this very personal and highly entertaining account of his experiences over the past five decades reveals why."
– *New Scientist*

"*The Secret Lives of Bats* dispels the bad reputation of bats, long lurking in our imagination as creepy, somewhat scary creatures – the bit players in horror movies and gothic romance. Tuttle's innovative photography, adventurous spirit, and compelling words and science reveal bats for what they are: intelligent, social, and fascinating mammals. In short, Tuttle's fifty years of research and conservation commitment has turned our aversion into awe." – Kathy Moran, *National Geographic*

"Tuttle's recent attempts to photograph them in their natural habitat have led him through many hair-raising adventures, which he entertainingly chronicles. A page-turning memoir of curiosity about – and dedication to – a significant part of the natural world."
–*Kirkus Reviews*

"Tuttle comes across as a conservationist and scientist first, a combination of Indiana Jones and Mark Trail second. The book is an adventure tale whose reward is always bats."
– *The New Yorker*

"It's a terrific read that will give you a much deeper appreciation of the 1,300-odd types of bats we now know exist, and all the work – and joy – that goes into understanding and protecting them."
– *Huffington Post*

"*The Secret Lives of Bats* highlights the lifelong journey of the man who arguably has done more for the conservation of bats than anyone else on the planet. Filled with personal and professional stories and peppered liberally with scientific insights about bats, this book is a must-read for a fascinating glimpse into the lives of the world's most misunderstood mammals."
– John P. Hayes, Colorado State University

THE SECRET LIVES OF BATS

My Adventures with the World's Most Misunderstood Mammals

MERLIN TUTTLE

An Imprint of HarperCollins*Publishers*
Boston New York

First Mariner Books edition 2018

marinerbooks.com

Library of Congress Cataloging-in-Publication Data
Tuttle, Merlin D.
The secret lives of bats : my adventures with the world's most misunderstood mammals / Merlin Tuttle.
pages cm
Includes bibliographical references and index.
ISBN 978-0-544-38227-5 (hardcover) – ISBN 978-0-544-39043-0 (ebook)
ISBN 978-0-544-81559-9 (pbk.)
1. Bats. 2. Bats–Research–Anecdotes. 3. Tuttle, Merlin D. I. Title.
QL737.C5T896 2015
569'.4074 – dc23
2015017314

Book design by Greta D. Sibley

Printed in the United States of America
24 25 26 27 28 LBC 11 10 9 8 7

I dedicate this book to the memory of
Verne and Marion Read,
who sponsored and participated in
many of my early adventures in
studying and conserving bats,
and to my wife, Paula, who
enthusiastically partners with me in
all my continuing activities.

CONTENTS

PREFACE

A City That Loves Bats

NEWSPAPER HEADLINES SCREAMED, "Bat colonies sink teeth into city." They claimed that hundreds of thousands of rabid bats were invading and attacking the citizens of Austin, Texas. It was September 23, 1984, and 1.5 million Brazilian free-tailed bats had begun moving into 16-inch-deep crevices beneath the Congress Avenue Bridge, just a few blocks from the State Capitol.

For more than a decade, Americans had been bombarded with scary newspaper and magazine stories that often included vividly imaginative accounts of rabid bats attacking people. "The Nightmare House," which appeared in the April 1, 1980, issue of *Family Circle*, even went so far as to claim that a family had been trapped in their home for three days and nights while a horde of bats attacked.

When Austin health officials added their warnings, many Austinites panicked and wanted the bats exterminated. Rather than viewing the great clouds of emerging bats as a wonder of nature, they were reminded of Alfred Hitchcock's terrifying classic, *The Birds*. Not surprisingly, people began imagining attacks from bats that were merely chasing flying insects, and Austin suddenly became the world center for scary bat stories.

Having just founded Bat Conservation International (BCI) to help people understand and appreciate bats, I felt that the new onslaught of frightening claims presented an extreme challenge. Nevertheless, I truly believed that Austin provided a golden opportunity to demonstrate that bats make wonderful neighbors. When I announced my resignation as curator of mammals at the Milwaukee Public Museum in Wisconsin so I could move BCI to Austin, nearly everyone thought I'd lost my mind. On learning of my plans, *Texas Monthly* magazine gave me its "Bum Steer" award, a tongue-in-cheek distinction given for especially embarrassing decisions.

With only one employee, I moved BCI to Austin in March 1986. We were just in time for the bats' spring return and within days, armed with several live ones, I began meeting with leading city officials, community leaders, and media representatives.

Fortunately, the truth about bats is extremely powerful. Free-tailed bats have strange faces, but their big brown eyes, winsome expressions, and gentle nature quickly converted countless skeptics. People frequently lost their fear in the blink of a bat's eye. When prominent socialite Roberta Crenshaw was informed of my efforts, she declared, "He might as well try saving cockroaches!" She assured her friends that if she had to view one of my bats close up she'd probably have nightmares for a month. Yet, when she finally gave in and took a look, she exclaimed, "Oh my – it's cute!" And she became one of America's first supporters of BCI and an outspoken promoter of Austin's bats.

Just two months after "awarding" me, *Texas Monthly* featured a story strongly supporting my bat conservation efforts. And just four years later, at a bridge-side ceremony attended by nearly 700 people, Mayor Lee Cooke proudly announced, "We've become the bat capital of America. It's really exciting that we've come from concern to a position where we want to protect, covet, and welcome the bats."

Even Governor Ann Richards became a bat fan. Soon after her election in 1991, she invited me to bring my bats to dinner to entertain her friends at the governor's mansion. It was an evening frequently punctuated by exclamations of "Wow – I had no idea! Did I hear you right? Bats from just one cave can eat more than a hundred tons of insects in a single night?" and "You really can train them like dogs?"

Time and again in my more than 50 years of studying bats, I've seen similarly dramatic conversions, perhaps explaining why I was brave enough to tackle Austin. It is simply amazing how quickly attitudes improve when people finally understand bats as they really are – sophisticated, beautiful, even cute, quite aside from their crucial roles as primary predators of insects, pollinators of flowers, and dispersers of seeds.

Austin now carefully protects its bats and has become the world center for stories documenting bats as safe neighbors and valued allies. In more than 30 years of sharing our city with the world's largest urban bat colony, no one has been harmed. And, simply left alone, these bats consume 15 tons of insects nightly. They also bring 12 million tourist dollars to downtown Austin each summer. On a typical August evening, people from all over the world line the Congress Avenue Bridge, jockeying shoulder to shoulder for favored views of the great emergence columns, truly one of our planet's premier wildlife spectacles. The bats pass within just a few feet of a flickering storm of flashing cameras, up to five columns at a time that can be seen for more than a mile as great black ribbons against sunset skies.

Of course, I can't share live bats close up – or even Austin's bridge bats – with more than a privileged few. However, in the following pages, I will share highlights from a lifetime of thrilling adventure and scientific discovery, covering every continent where bats live. From moonshiner standoffs to close encounters with tigers, cobras, and poachers – and bats as cute as any panda and as strange as any dinosaur, tiny bumblebee bats to giant flying foxes. Follow along, and I hope that through my adventures, you too will become passionate about bats.

ACKNOWLEDGMENTS

COUNTLESS INDIVIDUALS have provided critical inspiration and guidance, beginning with my parents, Horace and June Tuttle, who instilled a love of nature and can-do attitude and enthusiastically supported my early interest in becoming a bat biologist. Ernest S. Booth's talk to my fifth-grade class inspired me to become a mammalogist. Charles O. Handley Jr. provided early inspiration to study bats and later hired me to lead his Mammals of the Smithsonian Venezuelan Project. He also played an important role in getting me accepted into a top graduate program at the University of Kansas. Joseph C. Howell and Wayne H. Davis enthusiastically assisted me with my first published research. J. Knox Jones Jr. and Robert S. Hoffmann provided invaluable counsel as my major professors in graduate school, and my former wife, Diane Stevenson, taught me to write.

While I was employed as curator of mammals at the Milwaukee Public Museum, Verne and Marion Read and their family virtually adopted me, providing crucial encouragement and financial support, including Verne's participation as Bat Conservation International's (BCI) founding trustee. Many of BCI's greatest achievements would not have been possible without the Reads' leadership contributions.

Gordon M. Sears, chairman of T. J. Ross and Associates, Inc., and Bill Walker, president and CEO of Bacardi Limited, provided invaluable leadership guidance during the founding of BCI and served for many years as trustees. Additional trustees who provided special assistance include Jeff Acopian, Mark Adkins, Gene Ames Jr., David Bamberger, Eugenio Clariond, Mike Cook, Sharon Forsyth, Bob Gerrie, Don Grantges, Elizabeth Jones, Joan Kelleher, Travis and Bettina Mathis, Scott McVay, John Mitchell, Peggy Phillips, Mark Ritter, Beth Robertson, Andy Sansom, Lee Schmitt, D. J. Sibley, Marshall and Patsy Steves, Dave Weaver, and Mark Weinberger.

Most of my greatest achievements would not have been possible without the loyal support of hundreds of colleagues from around the world. They lent invaluable advice and credibility, shared their latest discoveries, partnered on projects, and provided articles for BCI's *BATS* magazine. Foremost among these were Luis Aguirre, Robert Barclay, John Bowles, Mark Brigham, Cal Butchkoski, Richard Clawson, Denny Constantine, Paul Cox, Bob Currie, Brock Fenton, Ted Fleming, Stephen Frantz, Steve Goodman, Leslie Hall, Daniel Hargreaves, Gareth Jones, Tom Kunz, Richard LaVal, Gary McCracken, Rodrigo Medellin, Shahroukh Mistry, Arnulfo Moreno, Scott Mori, Rachel Page, Lars Pettersson, Paul Racey, Fiona Reid, Ralph Simon, Fred Stabler, Peter Taylor, Don Thomas, Gary Wiles, and Don Wilson. I especially appreciate my longtime friendship with Mike Ryan, his early support of BCI, and the adventures we've shared while coauthoring 13 scientific publications.

BCI staff were additionally essential, especially Ed Arnett, Mari Murphy Houghton, Jim Kennedy, Amy McCartney, Linda

Moore, and Janet Tyburec. And BCI could not have succeeded in its mission without thousands of extremely generous members and donors and hundreds of partnering organizations, agencies, and corporations. Among BCI donors, in addition to the Read family, I owe special debts of gratitude for timely contributions from Ruth Adomeit; Luis Bacardi; Lee, Ramona, and Ed Bass; Bill and Carole Haber; Bettina and Travis Mathis; John Mitchell; Paxson Offield; Kenneth Olson; and the Brown, Lennox, and Kronkosky Foundations. The Nature Conservancy, the Tennessee Valley Authority, the American Cave Conservation Association, and the U.S. Fish and Wildlife Service were leading partners in protecting key bat caves, and the U.S. Bureau of Land Management, the U.S. Forest Service, the U.S. National Park Service, and the U.S. Natural Resources Conservation Service, and a long list of mining companies helped ensure the establishment of hundreds of bat sanctuaries that now protect millions of bats in old mines.

I am especially grateful to Mary Smith and many others at the National Geographic Society for critical advice and research support, for generous promotion of BCI in its early days, and for having educated millions of people worldwide about bats through publication of my five articles in their magazine. Also, Survival Anglia, a longtime leader in wildlife film productions, produced a documentary, *The Secret World of Bats,* that premiered nationally on CBS television in the United States and eventually aired in more than 100 additional countries. The film had a major impact in promoting public understanding of bats, and all who care about bats are indebted to cameraman-producer Dieter Plage.

My literary agent, Russell Galen, provided critical early advice and encouragement without which this book would not have been written, and my editor, Lisa White, made numerous insightful suggestions that greatly improved the final product. Jerry Hughes patiently led me into the modern world of digital photography after decades of reliance on film.

I've been given the priceless privilege of a life filled with discovery and adventure on behalf of an extraordinarily worthy cause that had been written off as hopeless. Mere words are inadequate to express the depth of my appreciation to all who have helped make this possible.

CHAPTER 1

TEENAGE DISCOVERIES

I'VE ALWAYS BEEN FASCINATED BY NATURE, so when, at age 17, I discovered thousands of gray bats, now referred to as gray myotis, doing things that, according to books of the day, they weren't supposed to do, I was immediately intrigued. It all began in April 1959 when a high school acquaintance told me about a bat cave near my home west of Knoxville, Tennessee. Baloney Cave was named for its baloney-shaped formations, and it was said to sometimes shelter thousands of bats.

The next weekend, I easily persuaded my father, who was always open for new adventure, to help me find it. We headed out on a beautiful spring afternoon. The sun was bright and the air was scented with honeysuckle blossoms as we followed a barely visible trail along a fence, then into the shade of stately old oak and hickory trees. A half-mile later, we found ourselves staring into a gaping pit about 12 feet in diameter at the top, sloping down like an antlion funnel. Limestone walls adorned in moss and ferns dripped from recent showers. This clearly was the cave my friend had described.

Wondering if the bats would still be there, we carefully climbed down into the cooler entrance, jumping the last few feet to the floor.

Before venturing into the dark interior, we retrieved our new miner's caps and carbide lamps from our knapsacks and added fuel. Each lamp included an upper and a lower chamber. We added quarter-inch chunks of carbide into the lower ones and poured water into the upper ones. When water contacts carbide it produces acetylene gas. And when the gas exits through a tiny nozzle in the middle of a shiny, metal reflector, it can be lit with a spark from an embedded flint. This provided each of us with a half-inch flame for light. We could alter the brightness by adjusting a lever, which controlled the rate at which water dripped onto the carbide. Even at their brightest, these lamps were dim compared to today's LED lights, but they were the best we had.

After allowing our eyes to adjust to the yellowish glow of our lamps, we began to look around, first noticing a room the size of a small bedroom on our left. It was strewn with old moonshine still paraphernalia, broken Mason jars, and parts of wooden barrels. The ceiling was smoke-blackened from the distilling process. Far more concerned about finding bats, we would later regret having assumed that moonshine stills in Baloney Cave were limited to the far distant past.

This was our first venture into a cave. My father led the way, as we stepped carefully around slick spots on an uneven floor, our hands often supporting us against the moist limestone walls. After going by several side passages, my father exclaimed, "Wow! Look at this." We were just entering a room the size of a two-car garage, which our dim lights barely covered. Along one side, baloney-shaped formations ran down a wall into a pit; because the bottom was beyond the reach of our lights, it seemed endlessly deep.

"I sure hope the bats don't live beyond that," I commented, pointing into the chasm.

So far, we hadn't seen any bats, though the cave floor was often strewn with a soft, dry accumulation of old droppings, commonly called guano. Although we were a little nervous about becoming lost, we decided to search the side passages. Some were quite narrow, requiring tight squeezes. My father finally suggested we return to the entrance and try to follow the most recent bat droppings like a proverbial trail of bread crumbs. As we were beginning again, we noticed a big difference between bat and rodent droppings.

Bat droppings were similar to mouse pellets, about a quarter-inch long by less than an eighth-inch wide. However, unlike those of deer mice and pack rats, which also frequent caves, once bat droppings dried, they easily crumbled into dozens of tiny fragments that reflected our lamps' light in a rainbow of colors. The reflections came from tiny bits of insects the bats had eaten. Rodent droppings were composed mostly of plant material; they were hardened like bits of gravel and failed to reflect light.

By following only the freshest trail of evidence, we were able to move ahead with much greater confidence. A few minutes later, through a small hole in a wall, I heard my first chittering bats.

On the far side of a small room, I could barely make out a furry mat of several thousand gray myotis covering the ceiling. They apparently had heard us coming and were rapidly waking up. Clutching my butterfly net in one hand, I desperately attempted to scramble through the hole in time to catch a few before they could escape, but to no avail.

Suddenly the air was virtually saturated with flying bats. Dozens were landing on my head and shoulders, because I was inadvertently blocking their escape. They were crawling down my neck and into my shirtsleeves – no need for a net! But I soon realized that they meant no harm and were only seeking places to hide. In fact, they had far more to fear than I did.

This was my first lesson in the gentle nature of bats, especially those that form large colonies. They neither scratched nor bit me as they swarmed over me, though I did have to hold quite still in order to avoid inadvertently crushing them. When the pandemonium finally cleared, my father helped me retreat from the hole clutching a couple of struggling specimens, which we gently placed in a cloth bag.

Back at the cave entrance, we got out my father's *Peterson Field Guide to Mammals* to identify our bats. Based on their unicolor gray fur, we easily identified them as gray myotis (*Myotis grisescens*). All other cave-dwelling bats of North America have bi- or tricolored fur – that is, each hair has a dark base and lighter tip or is banded dark, light, and darker from base to tip.

Curious to see how many bats would emerge to feed that evening, we waited in the cave entrance. The first individuals left about a half-hour after sunset, followed by a steady stream of thousands. We sat quietly, listening to the flutter of wings, which sometimes passed within inches of our faces. I was enthralled. Where were they going, and when would they return?

Several nights later, when we brought my mother back to see these bats emerge, we had a big disappointment. The few books of the time that mentioned gray myotis stated that they lived in a single cave year-round. Nevertheless, ours had disappeared. Speculating that perhaps gray myotis just didn't come out every night, we returned repeatedly to check their roosting area and watch for emergences at the entrance. But they were gone. They returned in September, then mysteriously disappeared again. Had they gone somewhere else for the winter? I was determined to find out.

With help from my father and some high school friends, I thoroughly explored even the innermost reaches of Baloney Cave on the chance that our bats had simply found a remote area where they could hibernate undisturbed. But we found no sign of them.

When our bats briefly returned and left again the following spring, I was convinced they must be migrating, despite books I'd read that said they didn't migrate. Armed with field notes, documenting when the bats were present versus absent, and two museum-type voucher specimens to substantiate my identification, I convinced my mother to drive me to Washington, D.C., so I could meet with scientists at the Smithsonian's Division of Mammals.

Blindly hoping we would be able to find scientists who could help, we approached the Smithsonian reception desk and announced we had bat specimens we'd like to discuss with their mammalogists. The receptionist calmly called Dr. Charles Handley, who headed the Division of Mammals, and a few minutes later he met us at the elevator. Dr. Handley, a bat specialist who eventually would become my mentor in a career as a professional bat biologist, had received guidance at an early age himself and was delighted to help me. Before we left, my mother and I had been taken on a grand tour of their vast collections and invited to come back any time.

He also introduced us to John Paradiso, the person in charge of bat banding for the U.S. Fish and Wildlife Service Bird and Mammal Laboratories. Impressed with my field notes and specimens, he encouraged me to take several thousand bat bands to see if I could track gray myotis movements.

He explained, "Each band has a unique number and says 'Write USFWS, Wash. D.C.' If someone finds a bat with one of your bands on its arm, they may write and tell us where they found it, and we'll let you know."

The bands were sequentially numbered in lots of 100 and strung on fine wires. Each had to be carefully pried open, with the two lips exactly parallel, to avoid injuring wings when closing them around bat forearms. We would have to open our bands exactly an eighth of an inch, then restring them on a larger, plastic-coated electrical wire. Once captured, each bat would have a band slipped over one of its forearms. A bat's forearm is the same as ours except for the attachment of a wing membrane along the rear edge. That membrane connects the body to the arm and fingers to form a wing. When mostly closed, the band would not fall off but it would not pinch the bat's wing membrane.

I could hardly wait to get home to inform my father and begin planning how we would capture and band our bats. Before we could band any bats, though, I had to open and restring each band in numerical order. It was a time-consuming task that limited the number of bats we could band.

By early September, I had 200 bands opened and ready for use. My father had made me a special hand net with an extendable handle, and I was equipped with a new field notebook. I could hardly wait to get started. But time after time we found few, if any, gray myotis in Baloney Cave.

By late September, we were seriously disappointed. Did they come every year or perhaps only sometimes? In early October, following several cool nights, we reentered the cave and quickly headed for the bats' last known roosting site in the inner room. Faint squeaks from above alerted us that we were about to walk right by a small cluster of gray myotis. They covered only a little more than a square foot, so I was able to clamp my hand net to the ceiling around them without harming any. When the frightened bats attempted to fly, they all fell into my nylon net. This was perfect! We'd found all we were prepared to band, and they were sleepy enough to be caught.

Carrying our netted bats, we scrambled up to a flat rock just outside the entrance and began sorting them by sex. Bat sexes are easily distinguished, just as in most other mammals. There were 226 males and 12 females, of which we banded 199.

The next day we returned to find a similar number of bats in the same place, but we were surprised not to find a single banded individual. Had our bats somehow gotten rid of their bands, or had they left and been replaced by newly arriving bats? We wouldn't know the answer until we found some of our banded bats.

The odds of someone else finding one of our banded bats and actually reporting it to the Fish and Wildlife Service were slim, based on the prior experiences of others – less than one in a thousand. However, Dr. Wayne Davis, a bat-banding biologist in Kentucky who would soon coauthor the book *Bats of America*, encouraged me in the possibility of finding some of my own banded bats. I had become acquainted with him through Paradiso at the Smithsonian, and he too became very helpful.

The best hope seemed to be to find our own bats, but only if they hadn't gone too far. For several months, my father and I had been asking neighbors, especially old-timers, if they knew of possible bat caves in the area. We'd heard plenty of wild tales about caves surrounded by endless mysteries, most not involving bats. In those days, people thought us crazy to be looking for bats, especially since we didn't seem interested in killing them. Even when we did hear about real bat caves and went to check them, we mostly found just sticks and stones, evidence of people's frequent efforts to kill bats at their easily recognized roosts.

When active gray myotis roost on limestone cave ceilings, their tiny claws, combined with the acidic action of carbon dioxide from their breath, etch the soft stone. Long-term contact stains the rock a rusty red color. The etched and stained roosts remain recognizable for many years after bats have left or been extirpated.

Even from our earliest observations, it was obvious that gray myotis had declined alarmingly. To find them would require exploring caves or parts of caves that humans seldom visited, often because they were remote or dangerous to enter.

One old man's story particularly caught our attention. We found him cooling off on a hot August weekend in a small country store. When he heard us asking the owner about caves with bats, he came over with an incredible tale. "When I's just a young'un, my buddy an' me sneaked inta a cave what people thereabouts was afeerd of. Could drop rocks in an opening and hear 'em goin' deep, deep down. The doc what owned it tried ta keep people away, afeerd they'd fall in and be bad hurt. We was scaret too, but we did sneak in fer 'nuf ta see it be plum full o' bats. Them suckas was coverin' ever'thing, like swarms o' bees. Near scaret us ta death!"

He went on to explain that the cave was located in a remote hollow between Clinch Mountain and Kyles Ford, a bit more than 100 miles northeast of Knoxville. "Just ask for Doc Pearson's place. Ever'one knows 'im."

Doc Pearson, the only doctor for miles around, was easy to find. He and his wife, Kate, quickly confirmed that their cave contained lots of bats. Prior to the arrival of electricity, Kate had relied on one of its cool sinkhole entrances to store milk and cheese. They had never entered the actual cave because of its difficult access. One entrance was a sheer 50-foot, freefall drop to a rugged floor, and the other two were narrow crawlways that sloped precipitously. Despite never having entered, Kate had often been surrounded by emerging or returning bats as she did her chores and knew better than to be afraid. She and her husband were quite curious to learn more. I guess they trusted me not to get hurt, given that my six-foot-four-inch-tall, muscular father looked quite capable of keeping me safe. He also exuded confidence, even though neither one of us actually knew much about surviving in caves yet.

Their cave was located in an oasis of natural wonder in the middle of an otherwise average cow pasture. The shaded sinkhole was lined with moss- and fern-covered boulders, surrounded by a small patch of forest where wild raspberry vines and sassafras plants grew. The cave's three entrances connected into a large main room, and a series of complex passages provided warm and cold air traps and stable conditions year-round, ideal for hibernating bats.

Bats are among the few true hibernators. The breathing of a hibernating bat is imperceptible. Its heartbeat drops from roughly 400 beats per minute when awake to about 25 in hibernation. Its body temperature often falls to within a tenth of a degree of surrounding cave walls. Some species can even survive subfreezing body temperatures. Most cannot. Each has its own unique hibernation strategy, and because of its complexity, Pearson Cave could accommodate at least eight species.

For a budding bat man, this was a magical place. We first entered in early December. My father and I had to slither through an icy crevice into a large chamber 60 feet wide by 150 feet long. Almost immediately we began seeing hibernating big brown bats (*Eptesicus fuscus*), surrounded by icicles. Rafinesque's big-eared bats (*Corynorhinus rafinesquii*) hung higher up where they were safe from freezing, their huge ears curled like a ram's horns. And a little farther in, we discovered a wall covered in an estimated 11,000 gray myotis, including a dozen with bands on their forearms.

I carefully wrote down the numbers so I could report them.

I had never before seen more than a dozen or so hibernating bats, much less such an array of species. On a moist wall, near a stream, we found scattered little brown myotis (*Myotis lucifugus*), then tiny tri-colored bats (*Perimyotis subflavus*), beaded in so much moisture condensation that they resembled Christmas ornaments. Aside from their moisture droplet disguise, tri-colored bats are easily distinguished from all other cave-dwellers by their distinctly tri-colored, yellowish fur and orange forearms. Unlike gray myotis and big-eared bats, they spend summers roosting in tree foliage, where their fur color conceals them like dead leaves.

As my father and I drove home later that evening, I kept exclaiming, "Wow! Can you believe we found so many bats? I wonder how many more there are?" And my father reminded me, "We need to find out who else is banding gray myotis and where theirs came from."

Once home, I made a beeline to retrieve my banding notes to see how close this other person's band numbers were to mine. I had never suspected the bands could be mine – after all, even if my banded bats did migrate, shouldn't they have gone south for the winter? – so I was astonished to find that they were my own numbers. I ran through the house yelling, "Can you believe it? We found our own bats!"

Not yet aware of the harm that could be done by disturbing bats too often during hibernation, we returned a week later to share our discovery with my mother, who also was a nature enthusiast. The Pearsons were fascinated to learn that at least some of their bats had come from as far away as Knoxville, but all I could think of was how many more of our banded bats we might find.

My mother wasn't enthusiastic about the icy slither to enter, but she was a good sport. We were disappointed to find that all

11,000 of the gray myotis we'd told her about had departed, apparently seeking a more secure hiding place deeper inside. We did find a few hundred gray myotis, but most eluded us.

We decided to climb a cedar log some previous visitor had propped against a wall in order to reach a side passage 12 feet up one side of the main entry room. My father steadied it as I, and then my mother, climbed, using nubs of cut-off branches like rungs in a ladder. Then, against my mother's protests, he climbed it alone. We found several thousand gray myotis high above the floor in this passage, but not as many as we'd hoped.

I suggested we follow bat droppings down several side passages, but the search was fruitless. Then, to save time, we decided one of us would go ahead to search and report. Since I could scarcely contain my need to explore, and my mother wasn't enthusiastic about crawling down small passages, my parents reluctantly agreed to send me. This was a mistake they would soon regret.

Eager for discovery and fearless, as teenagers are, I crawled rapidly along the next passage, leaving my parents to await my return. In my enthusiasm, I failed to notice how narrow it was becoming until my arms were pinned in front of me, and I could no longer back up. I had no option but to force my body ahead, hoping for a place wide enough to turn around. Unfortunately, the passage only got tighter as I squirmed with all my might, chilled to the bone and on the verge of panic. Minutes seemed like hours as I struggled forward. Then, just as I had nearly given up hope of an escape, I felt expanding space ahead and realized I was emerging back into the main entry room.

Such a narrow escape should have sobered me, but at that moment I looked up and noticed that bats we had disturbed were flying high overhead and appeared to be disappearing into a solid wall.

Obviously, if there was a passage invisible to humans, it could lead to the bats' secret hiding place. Throwing caution to the wind, I forgot about my concerned parents still waiting far behind, and began climbing fissures in the cave wall to get a better look. Sure enough, the bats were disappearing into a narrow opening that led to a 75-foot-tall passage, a type referred to by cave explorers as a "canyon passage." The bats had led me to a major discovery, a part of the cave exhibiting no evidence of prior visitation. Once I entered, no one would have the faintest idea where I had gone. Nevertheless, I was too hot on the trail of discovery to turn back. I needed to follow the bats while I had the opportunity.

Soon, I found myself some 50 to 60 feet above the passage floor, having increasing difficulty bracing across the widening span. Relying on a carbide lamp, attached by a metal clip to the front of a baseball cap–type hat, I saw more and more flying bats, all going in the same direction.

At that point I came to a portion of the passage where it was too wide for me to continue bracing between the walls. Despite being high above the floor, I was too excited to stop. Spotting a horizontal crack along one side, I inserted both hands, allowing my body to hang free, and continued hand over hand for several feet until the walls narrowed again.

My heart pounding, I found a hole in the wall from which I could hear bats, lots of them. I poked my head in, and there was the mother lode!

For as far as I could see, the walls were covered by a dense, furry mat of approximately 100,000 gray myotis. Each about two and a half inches long, they were packed in at more than 200 per square foot. Only the nose, ears, and wrists of each bat were visible. Little did I know I had discovered part of one of America's largest populations of hibernating bats. And at a glance I could see several more of my shiny aluminum bands.

Squeezing feet-first through a tiny opening in the ceiling, I dropped several feet to the floor without even considering how I might climb out. The bats must have heard me coming, because many were waking up and beginning to fly as I frantically grabbed the last banded ones before they could escape.

As soon as I had the banded bats in my mesh bag, I retreated into a small side passage to record their numbers. Before I could finish, however, the flame in my carbide lamp died, leaving me alone in inky black darkness. The only sounds were those of dripping water and the occasional flutter of bat wings. Still, I wasn't yet fully aware of the gravity of my situation. The first time my lamp went out, I spat into the water container and promptly got it started again. I mistakenly concluded that a flying bat had run into it, never considering the possibility that I'd run out of fuel. So I resumed writing notes. It wasn't until minutes later, when it went out again, that I realized I was in serious trouble.

I had violated all good caving rules – no hard hat, no spare light, no extra fuel, and my parents had no idea which direction I'd gone.

Shivering in damp clothing and a 43-degree breeze, I sat in total darkness on a guano-covered limestone rock, figuring I'd soon be hypothermic. But retracing my route without a light wasn't an option, and given my location in a previously undiscovered part of the cave, far from where I'd last been seen, there was little hope of rescue. Deciding that I was doomed to die, I laid my small notebook of band records on a ledge and sat, chilled to the bone, desperately searching for unspent fuel.

I carefully sifted the spent carbide powder into my left hand, combing through it for any small lumps that might remain. There were none—just the fine powder from used-up carbide. Nevertheless, I poured the powder back into the lamp in almost superstitious desperation.

I sat for what seemed like an eternity; then, for no sensible reason, I tried again to light my lamp. To my amazement, there was a sputter, and a pinhead-sized blue flame appeared. It didn't give enough illumination to see more than a foot or two, but I decided I'd rather risk falling to my death than give up to die from the cold.

I put my small notebook and now empty bat bag back in my pocket and began, in my mind, retracing exactly how I'd come into the small room with its hole in the ceiling. With the bats now gone, I found handholds to climb the wall and successfully exited the hole. Soon I came to the deep canyon passage, which I had to cross hand over hand, suspended far above the floor. Negotiating this difficult crossing had been terrifying even with a working headlamp, but with just a pinpoint of light remaining, it seemed suicidal. Each time I turned my face up to begin the crossing, my nub of light sputtered, threatening to go out.

After a long pause, I realized that death was certain if I didn't at least try. Time was running out. With my lamp barely flickering, I proceeded, just hoping that I had found the right crevice and that I'd find a foothold at the other end.

As I arrived on the far side, where I could again brace between walls that were closer together, my light failed. I'd made it past the greatest danger, but I was still without a light to help me find my way out. Carefully feeling my way along rough-surfaced walls, with only the fluttering sounds of bat wings to guide me, I eventually came to a location where I could hear my father calling in the distance. "Merlin, Merlin! Where are you?" I returned his calls, and we homed in on each other. Finally, at an opening too small for either of us to crawl through, he was able to hand me new carbide and water. Armed with a light, I found my way out, substantially extending my career in bat research.

I could hardly wait for spring. Would the same bats return to Baloney Cave? Sure enough, come spring, our banded bats from Baloney Cave came back, but again they disappeared within a few days. Now we had proof of round-trip migration, but we still had no idea where they had gone for the summer, nor why they had migrated north for winter to begin with. One thing was clear: to answer these questions, I needed to find a lot more bat caves, band lots more bats, and recruit lots of friends to help.

One hurdle that had to be overcome was the tedium of opening bands. Fortunately, my father was a natural inventor who soon developed a simple tool for quickly opening and stringing bands. His tapered and polished steel rod had a bulge in the middle and a detachable wooden handle at the back end. Groups of closed bands could be pushed along and pried open an eighth of an inch at the bulge, using a round piece of steel with a hole in it to push them along. Once opened at the bulge, the bands slid down the reverse taper onto a larger-diameter wire. This gadget enabled me to open hundreds of bands per hour instead of merely dozens. It also kept them in numerical order, a critical improvement.

Because I was looking for other caves, after the spring of 1961, Baloney Cave was no longer a focus for my efforts. However, when I returned that fall with my father's biology class from the nearby Little Creek Academy, we were in for a big surprise.

Everyone waited outside while I quietly entered to check for bats. But, as I jumped down the entrance drop to the cave floor and ducked inside, I realized I wasn't alone. Just a few steps into the half-light near the entrance, I saw two things: a sawed-off shotgun pointed at my skinny midriff and the face of a stranger with just one good eye. The bad eye was a mess of scarring and empty socket.

I didn't know who this was or what to say to him. "Are there any bats here today?" I asked when the silence lasted more than a few seconds. "No," he grunted. "This is my cave. Who else is with you?" Clearly, he had heard voices outside and was not at all interested in my banded bats.

Before I could answer him my father called down. "Merlin, what's going on?"

"Come see," I said.

The man in the well-worn shirt and trousers pointed his shotgun up toward the sunlight as my father scrambled down into the cave. He didn't seem as shocked as I imagined he would be to find me facing a scruffy old man with a shotgun. We had been living in this rural area near Knoxville for only a couple of years and knew little about moonshiners.

"Hello, nice to meet you," Horace Tuttle calmly greeted the stranger. "I teach biology, and my son here and some of my students outside just want to see the bats. We don't mean to bother you, sir. Do you mind if we just have a quick look and be on our way?"

Another long silence from the one-eyed fellow as he stared at us. "Okay, but do it fast. Jes' you twos. And don't be comin' back again."

We lit our carbide lamps and walked some 50 feet farther in, pretending to look for bats on the ceiling and walls while the man and his shotgun followed us several paces behind.

Near the entrance, as we passed the side chamber, the edge of our light beams caught a motley collection of copper tubing, a small wood-burning stove, barrels, and Mason jars. Overwhelmed by the strong odor of fermentation, we quickly aimed our lights away and moved on. Without continuing far enough to find any bats, we turned around, ready to leave. The man behind us must have known we'd seen his still.

Turning around to face him, my father said in a matter-of-fact tone, "Thanks for letting us look for bats in your cave. We'll be on our way now." The man said nothing, remaining inside the cave as we left with the biology class.

On the drive home, I rode with my father, a conservative and religious man who did not use liquor or tobacco. He was direct: "What we saw is illegal. We'll have to call the sheriff." We were naive about moonshiners and the lengths they could go to in protecting their source of income, not unlike drug dealers of today.

Soon after returning home my father phoned the Knoxville sheriff's office to report an illegal still in Baloney Cave. He described the moonshiner as having only one good eye. The sheriff said he must be Bad-Eye Murphy. He'd lost an eye in a shootout with sheriffs.

With an agreement that we would remain anonymous, my father told the sheriff we would take him to the still. Well after dark that evening we were surprised when half a dozen carloads of officers arrived at our house. We'd expected just one or two vehicles, but the sheriff wasn't taking any chances.

I dutifully joined my father in leading the half-mile hike from the nearest road to the cave. The sheriff and his deputies were afraid we'd be spotted and shot if they used lights, so we had to advance at a snail's pace, guided only by faint moonlight. On arrival, the deputies formed a semicircle around the cave entrance, hiding behind trees and rocks. They then yelled something along the classic lines of "Come on out. We've got you covered."

When there was no response, they shone a spotlight into the entrance while the deputies cautiously advanced from two sides.

To their disappointment, Bad-Eye Murphy was not to be seen that night. The officers destroyed the still and we went home, thinking that would be the end of it.

The next morning, the *Knoxville Sentinel* carried a headline proclaiming "Tuttle and Biology Class Lead Sheriffs to Biggest Raid of Year." Much to the horror of our family and neighbors, the story named not only my father but also the school where he taught. Everyone assumed retribution would come. Our neighbors helped set up floodlights around our home, but that was about the extent of the protection they gave us.

My uncle, Grant Tuttle, had lived in the area longer than we had and knew how locals talked to each other. He believed the only solution was to appeal directly to Bad-Eye Murphy.

He found out where the moonshiner lived, loaded up some food gifts from the church school's garden, and drove our family out to apologize. Murphy lived at the end of a long dirt road in a heavily wooded area. As in the cave, it seemed like we had barely turned in his direction when he stepped out in front of us. Deer rifle in hand, he ordered us to stop. How he happened to be waiting, I'll never know.

My uncle had the gift of gab and immediately stepped out of our station wagon and began apologizing. "No need for your gun, friend. This is just my family here. No sheriffs. We've come to say how sorry we are the way everything happened.

"My brother, he just moved to this state from Florida. He has no experience at all with whiskey making. He stumbled into your business when he was just trying to be a good teacher and father to his young son Merlin here, who is interested in bats. Merlin's got more knowledge of bats than any young'un I've heard of. No one intended to make trouble for you. We've got some nice pickings from our garden for you and some preserves."

Uncle Grant sure could schmooze.

"You know how some people get carried away with what they think the Lord would want? That's what happened. My brother is a God-fearing man and he believed he was doing the right thing by calling the sheriff."

Murphy seemed genuinely dumbfounded that we'd dare come to him to make amends. After several minutes of my uncle's explanations and promises that no such mistakes would be repeated, Murphy softened.

"You're lucky you done the right thing here," the moonshiner said. "Something bad could've happened to you if I believed you were still out to get me. You know I'm jes' tryin' to make a livin' for my family."

My father had one simple response: "Thank you, Mr. Murphy."

Bad-Eye Murphy turned toward his house. My uncle got back in our car, and we quickly left. We didn't talk a whole lot on the drive home. I realized what my father had done was not a smart thing, but I didn't dare tell him so.

That wouldn't be my last run-in with armed locals. My gray myotis research would take me to more caves in places where backwoodsmen didn't know anything about young scientists. I was simply a stranger to them and their habits of distilling and drinking what they called white lightnin'.

I didn't yet have any idea that my teenage discoveries at Baloney and Pearson Caves and my resulting curiosity would lead me to band and track 41,000 gray myotis across a dozen states.

CHAPTER 2

SAVING THE GRAY MYOTIS

WHEN I ENTERED GRADUATE SCHOOL in 1968 to obtain a Ph.D. in biology at the University of Kansas, I hadn't lost my curiosity about gray myotis. Where did they go, and why did they require different caves as the seasons changed? Would any of the bats I had banded nearly a decade earlier still be alive?

I decided to answer such questions as part of my research for a dissertation. It was only natural that my initial fieldwork centered on Pearson Cave. The doc had died, but his wife, Kate, was still protecting the cave and was delighted to see me.

To better understand the annual activity cycle and needs of gray myotis, I set a bat trap in the most-used entrance for one night every 12 days for the entire year of 1969–1970. It wasn't long before a couple of men I suspected of owning stills in the area came by to check me out. I explained my research and invited them to watch the trapping that evening.

My trap consisted of two six-by-five-foot aluminum frames with hundreds of vertical fishing lines strung between them. It looked like a harp, but with adjustable legs to support it a few feet off the ground and a canvas bag hanging below to hold captured bats. This device, which I had recently invented, was capable of catching thousands of bats per night, enabling me to sample and release large numbers without harming them.

Several moonshiners and their friends showed up that evening to watch me trap bats. They were fascinated by the contraption that caught these creatures as they darted out on ten-inch wingspans. As I plucked the bats from the trap for examination, I explained to them that a single bat could catch a thousand insects in just one hour and how much that added up to when tens of thousands went out to feed.

Meeting a scientist on their home turf intrigued the moonshiners, who weren't shy about defending their profession. They were back the next night and soon began to treat me like some kind of mascot, calling me their bat man. They explained how their particular livelihood had gone on for generations and that the government was unfairly trying to tax poor people. Revenuers were viewed as despicable invaders.

I became especially good friends with one moonshiner, Hue Kyle Cellars, who was in his late twenties like me. He could neither read nor write, but he was a sharp-witted, thoroughly fascinating conversationalist.

After getting to know me, Hue Kyle invited me to have dinner with his family one night. He, his wife, Marjorie, and their eight children lived in an unpainted wooden shack with just a living room, kitchen, and two bedrooms. It was in such a state of disrepair that we couldn't all enter the kitchen at once for fear the floor would collapse.

They had prepared a smorgasbord of food just for me. I was touched by how this impoverished family had worked so hard to please me. As winter progressed and I continued my research visits, they insisted I stay with them. When I agreed, I didn't realize the hardship I would be causing. All eight kids, who normally shared one large bed, had to sleep on the living room floor on the nights that I stayed.

Once Hue Kyle and his partners learned my routine, I could just about count on hearing someone call an alert, "The bat man's a-comin'!" each time I entered the valley's single-lane dirt road. And, nearly always, they'd invite me to stop and have a "snort" of their craft. Their lightnin' was nearly pure alcohol and burned like fire, so I just barely touched it to my lips to be polite.

Hue Kyle turned out to be one of the most loyal friends I ever had. Our friendship lasted until he died of cirrhosis of the liver some four decades later. Of course I couldn't condone his being inebriated much of the time, what that did to his family, nor his shootouts with law enforcement officers. My eventual discovery that he slept with a sawed-off shotgun under his bed scared me out of continued overnight stays. I didn't want to risk being caught in a middle-of-the-night gun battle. However, having it known throughout the valley that he and his cronies considered me their friend made me feel secure when I was working late at night at Pearson Cave, in plain sight of the road.

It's difficult to describe how rough and lawless some of these backwoods hollows could be. The valley in which Pearson Cave was located could be entered only by a one-lane dirt road, and the general area was a well-known hangout for outlaws attempting to escape justice, just as in stories of the Old West.

As recently as 1989, when I returned with German filmmaker Dieter Plage, filming part of the documentary *The Secret World of Bats* for CBS television, we stopped by to see Hue Kyle. He was still living in his rundown shack. Dieter was anxious to see a real moonshine still, so I asked Hue Kyle to show us his. In the process, his uncle felt it necessary to get right in my face, sternly wagging a finger and warning me, "Jes' remember, ya double-cross Hue Kyle and ya gonna be dead suckas!"

Illustrative of just how risky working alone in some of my study areas could be, when a sheriff from Scottsville, in northern Alabama, found me banding bats late one night at Sauta Cave, he became concerned for my safety. He warned me, "This is a well-known hangout for dangerous characters." When I pointed out that the cave's colony of a quarter-million gray myotis was critical to my dissertation research, he sighed and said, "Okay, if you're determined to risk your neck, at least come by our office tomorrow and let us deputize you so you can carry a handgun." A bit unnerved by his warning, I politely thanked him and refused, fearing I might be at greater risk armed than unarmed. I'd already talked myself out of several potential jams, in part because I so obviously posed no threat. But a few nights later I had second thoughts.

I spent the evening of July 14, 1970, trapping and recording band numbers of emerging gray myotis at Gross Skeleton Cave, just a few miles from Sauta. It was a peaceful location with a crystal-clear stream and mossy limestone walls, hidden deep in the forest, away from prying eyes – or so I thought.

At midnight I removed my bat trap from the entrance, loaded up my equipment, checked my data, and began the relatively short drive back down a single-lane farm road. It was a hot, muggy night and I had the windows open, listening to the raucous din of thousands of cicadas.

Suddenly my reverie was shattered. A scruffy-looking man with a rifle in his hand was blocking the road ahead. Then a second similarly armed man emerged from the forest behind me. I had only a split second to think. The main highway was just a hundred feet to my left across a cotton field, so I ducked, gunned the engine of my Buick LeSabre, and swerved into the cotton field, hoping I could reach the highway without getting stuck or shot. Getting across that field seemed to take forever, but my trusty vehicle made it, so I'll never know what awaited me had I failed.

Most of the people I encountered in the course of my research were friendly, enjoyed hearing about my gray myotis discoveries, and sometimes were helpful even under the most surprising of circumstances. One night, when I was caught speeding in an Alabama radar trap, the officers couldn't resist inquiring about the strange aluminum frame strapped to my luggage carrier atop my car. Strung with fish lines, it resembled a large harp and even made a humming noise as I drove. "What is this thing you have on your car?"

I responded simply, "It's a bat trap."

Like most people, they said, "Oh, you mean a bass trap?"

"No, it's a trap for catching bats."

"Why would you want to catch bats?"

"I'm studying their migratory movements. I trap them at cave entrances, band them, and track their movements. So far, I've found bats from this area traveling to Florida, Kentucky, Virginia, and almost to Kansas."

"You're kidding – really?" Then, forgetting all about my speeding, they offered to guide me to a local bat cave. It wasn't the kind of cave where I normally would have expected to find my bats, as there was only about a foot of airspace between the top of the cave and a large spring. People were said to come from miles around to collect the crystal-clear water, which according to the officers was the best tasting in the state. They had noticed bats coming and going while patrolling this popular site at night.

The next day I returned and, barely able to keep my face above water, entered in search of gray myotis. I was quite surprised to find that the "especially tasty" spring water was flowing through a large pile of gray myotis guano less than 50 feet inside. Figuring that if people had been enjoying this water for generations without harm, there was no reason to scare them off now; I kept the secret to myself. However, I wasn't surprised to find some of the bats wearing my bands.

I couldn't resist getting a bit sentimental when recapturing my banded bats, some of which I hadn't seen since I was a teenager. Wherever I found them they were like old friends. It was always great to see them again, though I'm sure they weren't nearly as happy to see me.

Recapturing bats more than a decade after banding is no longer considered unusual, since subsequent research has shown unequivocally that bats are the world's longest-lived mammals for their size. In recent years, dozens of banded bats of several species have been recaptured 20 to 35 years later. In fact, a Brandt's myotis (*Myotis brandtii*) in Russia remained in excellent health when last seen at 41 years of age. Such survival for a bat requires excellent hearing, mobility, and coordination, the equivalent of a 100-year-old human still being able to run sprints through an obstacle course. Otherwise it would not be able to detect and chase flying insects and would starve. Certainly, scientists who study aging have much to learn from bats.

Ultimately, survivorship calculations based on many thousands of recaptures of my banded gray myotis projected maximum survival ranging from 10 years in some colonies to more than 40 in others.

As my discoveries became more interesting, I was able to recruit a growing number of enthusiastic volunteers, foremost of which were members of the Huntsville, Alabama, Grotto of the National Speleological Society—at least until they discovered what tortured lengths I would go to in pursuit of new discoveries. In Donal Myrick's 1972 book, titled *Fern Cave,* he says of my bat research trips, "He often went alone, but when he could, he would get some unsuspecting grotto member to accompany him."

Dave Weaver, perhaps the longest-suffering caver ever to assist me, many years later served as a trustee of Bat Conservation International. Shortly after his high school graduation in 1970, he volunteered his field assistance, traveling with me to dozens of caves, from Florida to Virginia. Dave was fearless, going along with whatever was required, roping off 100-foot cliffs, squeezing through tight passages, outrunning angry bulls, and respecting my moonshiner friends.

Many years later, reminiscing about his experiences with me, he laughed and said, "Merlin had a poor internal clock." Especially recalling a day spent searching for bat roosts in Fern Cave, Alabama, he said, "He would always lose track of time, saying we'd just be in a cave for a couple of hours. We'd come out twenty-four hours later, and we hadn't even eaten!" I in fact deny ever having kept anyone in a cave without food for more than 23 hours! That was in Fern Cave, the most complex bat cave I've ever seen. It was a real challenge. Gray myotis roosts were scattered across 12 levels and more than 15 miles of passages, including many that were 100 to 200 feet tall, requiring seemingly endless rappelling and climbing.

We would eventually discover that this one cave shelters the largest hibernating population of bats known worldwide. The cave is so complex that less than 10 percent of its roosting areas can be seen, even in a 23-hour trip.

The gray myotis in Fern Cave, a roughly estimated 1.5 million of them, were discovered by an alert member of the Huntsville Grotto, named Jim Johnston. One of the cave's original discoverers, Jim had been exploring other parts of the cave since 1961, before the bats were found.

Early in the summer of 1969, Jim and fellow grotto members were in a newly discovered part of this complex system when he spotted a single banded bat. Always enthusiastic about helping biologists, Jim caught it and reported one of my band numbers to the Fish and Wildlife Service.

When I was notified, I was ecstatic, not so much because a single banded bat had been found at a new location, but because of Jim's description of the cave where he had found it. He reported miles of huge passages that were exceptionally cold, with a thin layer of widely scattered bat guano on the floor.

I immediately called him, exclaiming, "I believe you've found the most important bat hibernation site ever discovered!"

He responded, "Didn't the Fish and Wildlife Service tell you that we found only one other bat and that we didn't find any major guano deposits? It's nothing compared to what we see in caves like Sauta."

I explained that hibernating bats leave their cold winter caves in spring and don't return until fall. Also, since they don't eat in winter, they do not leave significant guano deposits as they do in summer nursery caves.

I asked, "Can you take me there next weekend?" There was a long pause, and Jim asked how much training I'd had in vertical caving. He explained that just to enter the section of cave where he had found my banded bat would require us to rope off an initial 105-foot freefall, followed by another 75-foot drop, and that he couldn't take me until I'd had extensive training in vertical techniques.

"Please," I begged. "Take me anyway. I've worked in hundreds of caves, am in great physical condition, and can assure you I won't panic on a rope. You can teach me at the cave."

"Is it worth risking your neck just to see this place?' he asked.

My answer was "Definitely! And I'll be happy to sign a liability release."

Puzzled by my determination and extremely curious about my claims, Jim finally relented. So on July 21, 1969, he and his caving buddy John French, both experts who had pioneered early development of vertical caving gear, met me at a caver hangout near Huntsville for an early breakfast.

An hour later, we began a steep, mile-long hike up a rugged mountain, armed with about 60 pounds of rope and caving gear. These guys were very fit. Most people would have been exhausted just from climbing to the cave, loaded down with equipment.

As I was getting tired, we finally crested a ridge on the west side of Nat Mountain, where we stared into the gaping maw of a sinkhole about 80 feet in diameter. My guides quickly tied off a caving rope to a sturdy tree, threw the remainder over the sinkhole lip, and unpacked seat harnesses, brake-bar racks, and carabiners we would each require for controlled descent. Seat harnesses were made of automobile seatbelt material, sturdily sewn together to form 27-inch-diameter loops. The carabiners were made of oval-shaped steel, each with a lockable gate at one side. These would be used to connect seat harnesses to brake-bar racks.

We each would use a brake-bar rack through which the climbing rope would be looped back and forth around crossbars on a steel frame. The bars could be opened and closed to receive the climbing rope, then pushed close together to slow descent through friction or spread apart for greater speed.

Before looping the climbing rope through the brake-bar rack, we each had to put on a seat harness. Jim showed me how to wrap the upper part of a seat harness around my waist at normal belt height, then lower the bottom of the loop about ten inches and pull it forward between my legs. The ends were pulled together, one around each hip from the sides, and all three loops were then attached to a carabiner in front.

While Jim helped me, John quickly organized his gear and descended the relatively easy 50-foot drop into the sinkhole. I would go next, with Jim helping me rig at the top and John belaying me at the bottom. Jim explained, "That way, even if you lose control, by simply pulling on the rope, John can stop you."

Fortunately, my guides had brought an extra hard hat. Their confidence in my caving safety skills wasn't exactly enhanced when I admitted I'd never worn one before. It wasn't hard to understand the looks they exchanged. They obviously were wondering about the wisdom of taking a reckless novice into such a risky cave.

Jim attached a brake-bar rack to the same carabiner that held my seat harness together and explained how to open and close the brake bars so I could thread loops of the climbing rope through. The more loops around the brake bars, the slower the descent. He added four.

Once I was securely rigged, and Jim had made sure I was wearing good leather gloves to protect me against rope burns, he gave his final instructions. "Hold yourself upright to the rack with your left hand, and use your right to control the speed of your descent. To slow down, just pull on the rope below you. Even a slight pull with your right hand will stop you. However, no amount of pulling on the rope above is likely to slow you down. Keep that in mind. It's critical when you are no longer being belayed."

With John holding the rope below, I began slowly backing down the slope toward the drop-off. I soon realized that Jim had rigged me so tightly that I had to work just to go down. I stopped and complained, "This is supposed to be fun, not work." I convinced him to undo one loop around a brake bar and spread the remaining bars farther apart, reducing the friction. I could then descend much more easily, controlling speed with my right hand.

The real test came a few minutes later when we faced the 105-foot freefall drop into the Morgue entrance pit, so named because of the extraordinary number of large mammal skeletons at the bottom, some prehistoric. The drop-off was reached through a sharply sloping passage with a rubble-strewn floor. Fist-sized rocks rolled ahead, disappearing over a narrow ledge, clanging faintly far below. At that time of year, the bottom was obscured in dense fog, so we had to simply trust that our rope reached. John securely tied it to a log that had fallen into the sinkhole above while Jim tied a canvas pad in place to prevent our rope from being cut by the sharp-edged ledge.

Again, John descended first. When he had reached the bottom, detached his brake-bar rig, and stepped beneath an overhang that would help protect him against falling rocks, he yelled, "Off rope!" He would again hold the end of the rope in case I lost control.

"Your turn next," Jim said with a grin. He was still finding it hard to believe that I would not be chickening out by the time I got to the lip. I'll admit it was a bit scary going over the lip for the first time, still unable to see the bottom. But within a few feet of beginning the freefall drop, I was in love with vertical caving. I soon learned to plummet down, slowing myself at the last minute, not a highly recommended procedure, but lots of fun if you know your rope and equipment.

Once at the bottom, I quickly understood why I'd been warned to bring a coat. Entering sweaty from 90-degree outside heat, I felt as though we'd entered a walk-in freezer. Experienced with the problem and employed by the space program, then centered in Huntsville, Jim and John had brought along some of the first space blankets. They were originals, so thin and lightweight they could easily be carried in a hip pocket. And when we had to wait for each other to climb ropes, they were invaluable in keeping us from succumbing to hypothermia.

Jim led the way to a short crawlway that emerged into a tall canyon passage where I immediately began to recognize clear evidence of longtime use by large numbers of bats. Vertical walls were etched and lightly stained, similar to what I'd discovered as a teenager in Pearson Cave, and with such ideal cold air–trapping characteristics, I knew this would be one of the biggest discoveries of my career.

Hour after hour, we rigged ropes into new pits or braced between canyon passage walls, often 40 to 100 feet above floors. We searched a seemingly endless maze of options, but never lost sight of bat-roosting evidence. By the time we climbed out we were so tired we could barely walk back to our waiting vehicles. Nevertheless, we could hardly wait to see if my predictions of a huge hibernating bat population would be confirmed.

I'll never forget the suspense of our return trip the following January. This time I was accompanied by Jim Johnston, another of his caving buddies, Carl Craig, and three additional volunteers. Carl descended the entrance drop first while Jim made sure I was properly rigged. Then we had to wait for him and the other volunteers, in refrigerator-like conditions among chunks of fallen ice left over from the passage of the last cold front. Knowing that the first bats might be fewer than 100 feet farther in, I could barely resist the temptation to rush ahead. By the time everyone descended, it was hard to tell whether I was shivering more from the freezing conditions or from the hoped-for excitement of seeing the largest assemblage of hibernating bats I had ever encountered.

Seeing my excitement, my guides were happy to allow me to scramble ahead as we entered a narrow crawlway into the first suspected bat-hibernation area. The minute-long crawl seemed more like an hour, but when I could finally stand again, my headlamp illuminated huge clusters of gray myotis covering both walls of a tall canyon-like passage that stretched ahead as far as I could see. The bats were spread out, with loosely interlocking wings, covering thousands of square feet of walls. Gray myotis cluster at different densities during hibernation to help regulate body temperature. They can pack in at up to 300 per square foot in freezing temperatures or spread out to just 25 to 50 per square foot at 50 degrees. Thus they insulate themselves against further drops in temperature or expose greater surface area to aid in cooling when lower body temperatures can reduce metabolism and save energy.

As Jim and Carl led us, we roped down additional drops of 60 and 75 feet, crawled through tortuous side passages, and several times braced between walls high above floors. We had time to visit only a relatively small area of the cave, but we recaptured 387 of my banded bats that had come from 25 other caves across several states. These bats had migrated from as far away as Florida and northern Tennessee.

I have never seen more than a small fraction of this cave's hibernating gray myotis at any one time but have quite conservatively estimated that a million and a half spend their winters there. Though neither I nor any other scientist will ever find them all, it warms my heart to know they have such a refuge.

For me, the worst part of my work in Fern Cave was always the climb out at the end. First, I had to sit patiently for up to an hour in an icy draft of frigid air while associates rigged up and climbed. Then, chilled to the bone, it would finally be my turn. Though dropping in could be great fun, the climb out could be torture.

In the old days we lacked modern harnesses that now make climbing much easier. We relied only on our seat slings and a pair of Jumars. Jumars are Swiss-made mechanical devices designed for climbing a fixed rope. They are equipped with handholds and toothed cams that are hinged to grip the rope and easily slide up but lock when pulled down. The lower one was attached by a strong cord to a foot, the upper one to the seat sling. To ascend two feet of rope, I had to hang from the upper Jumar while pulling the lower one up. From a nearly horizontal position, I'd then push and pull my body to be vertical while advancing the upper Jumar. It was the equivalent of having to do seemingly endless squats and chin-ups at the end of an already exhausting day.

I have indeed led some 23-hour trips into Fern Cave, and it wasn't uncommon to run short of food and water. Nevertheless, we never visited more than a small fraction of the roosts or saw anywhere near all the hibernating bats. It is important to note that, because of Fern Cave's vast complexity, we could visit a sequence of locations, thus avoiding prolonged or repeated disturbance of the same roosts. Repeated disturbance can threaten bat survival by forcing them to waste precious energy that must last until spring.

I returned to Fern Cave many times in the course of a decade of study, usually accompanied by volunteers, but also alone when none were up to the challenge. On several occasions I emerged so exhausted that I seriously worried I'd be unable to make it back to my car. I probably owe my life to the dozen or so Huntsville Grotto volunteers who, despite my excesses, continued to help carry equipment, taught me safe caving techniques, and kept me from getting lost in that endless maze.

Ultimately, thanks to early support from my parents, combined with thousands of hours of volunteer assistance and colleague collaboration, I was able to track gray myotis movements across many thousands of square miles in ten states. From more than 23,000 recaptures of banded bats, I was able to answer my original questions and piece together the most important reasons for lengthy movements and curious behavior.

Baloney Cave turned out to be used as a migratory stopover roost, as I had originally hypothesized. It was too warm for winter hibernation and too cool for rearing young. But its intermediate temperature was good for torpor and fat storage, and it was conveniently located near the Tennessee River. Bats pausing in this cave were from several warmer caves where nursery colonies reared young.

In Pearson Cave, up to 100,000 gray myotis formed single, dense clusters at near-freezing temperatures. In this manner, they lowered metabolic rates to an absolute minimum, saving energy for longer migrations. These bats also saved energy during brief winter arousals from hibernation by occupying heat-trapping domes in the upper cave. Temperatures there could be up to 20 degrees warmer.

I learned that the longest-distance migrators often spent as much energy traveling during fall migration as during an entire winter of hibernation, explaining their need for extra energy savings. Gray myotis preparing for the longest migrations more than double their lean body weight in fat.

Given its ideal energy-saving conditions, Pearson Cave attracted gray myotis from the longest distances I was able to record. They came from all directions, including from Kentucky, Virginia, and North Carolina, west to central Tennessee and south to northern Florida. Many of these bats had to live off of stored fat reserves for up to six months annually.

Fern Cave provided exceptional protection from humans, but lacked the lowest, energy-saving temperatures of Pearson. Fern's greatest advantage was its central location near extra-warm nursery caves and rich, over-water feeding grounds. Bats that ate more and traveled less arrived with more fat and could emerge earlier in spring. Not surprisingly, my studies revealed that the largest gray myotis colonies were invariably found where maximum energy was saved by rearing young in extra-warm, heat-trapping caves located near rich, over-water feeding grounds and suitable hibernation caves.

Perhaps most importantly, I learned that no single colony is typical of its species. Each faces a unique combination of costs that must be balanced by savings in other areas. Caves that are shaped and located such that gray myotis can occupy them year-round are exceedingly rare. Nevertheless, a study at one such cave had led to the long-held misconception that this species was nonmigratory.

As would be expected, only nursery colonies in extra-warm roosts near rich feeding areas could rear young and put on sufficient fat quickly enough to afford the longest migrations, such as from Florida to northern Alabama or Tennessee. Conversely, those that had to rear young in cooler caves or at greater distances from feeding areas couldn't afford the cost of more than short migrations, and they required ideal conditions for hibernation.

Colonies typically used from two to six caves within a 30- to 50-mile-long area bordering a major river or lake. I defined this area as their home range.

During the period from late May through early July, when young were reared, colonies split into a nursery and one or more bachelor groups. Like most other bat species, each mother gave birth to a single pup. Bachelor groups were composed mostly of males and yearling or otherwise nonreproductive females. Gray myotis, especially adult females, were extremely loyal to the area of their birth, as well as to the hibernation cave chosen for their first winter.

Additionally, both sexes sometimes traveled far from their home areas, quite aside from their migratory movements. I banded an adult female, number 7-2638, and released her on July 8, 1969, at the entrance to her nursery roost in Harris Cave west of Decatur, Tennessee. I next captured her near Sauta Cave, 150 miles southwest of her home area, on August 7. She seemed a long way from home. But just four days later, I caught and released her again back at her home cave. I soon learned that both sexes sometimes travel long distances, apparently just exploring or visiting "friends."

I remain curious regarding how young gray myotis find and choose the hibernation cave to which they will return each winter for the rest of their lives. Colony members typically disperse among several such caves, which may be located in opposite directions. How, then, would new pups know which one to choose? They couldn't follow their mothers, because most mothers migrate at least two to four weeks ahead of their pups.

Though many adult males travel to hibernation caves, either with or ahead of females in order to get a head start on courting and mating, I never found juveniles completely abandoned.

Groups of young gray myotis were always accompanied by at least one or more adult males, giving me the impression that experienced males were serving as guides. If this were true, how would males decide which pups to guide and who should do it? Why would they want to give up extra sex for such an onerous task?

Since migratory stopover and hibernation caves often cannot be seen from more than a few feet away, and because migrant bats come from all directions, it is inconceivable that young bats could find such caves unaided. Recent behavioral observations by Chinese scientists suggest that bats can sense the earth's magnetic fields, but even the famed homing pigeon cannot find its way to a location where it has never been.

I was additionally puzzled by the fact that certain pairings or small groups of gray myotis were often found roosting or traveling together at locations many miles apart over periods of nearly a decade. It seemed that these bats formed long-lasting relationships.

In Europe, modern microchips, commonly referred to as PIT-tags, have recently been used to more precisely track bat movements and behavior. Gerald Kerth, Nicolas Perony, and Frank Schweitzer reported that similar-sized "Bechstein's bats are able to maintain individual relationships in a highly dynamic social environment over extended periods of time." In fact, they documented bat social structure resembling that of elephants, dolphins, and humans, including cooperative behavior such as information sharing. My experience with gray myotis indicates that they too are highly sophisticated. Even so, I still can't understand what kind of social sophistication could lead males to forgo sex for babysitting! Obviously, even now much remains to be learned.

One thing that I couldn't avoid seeing was that gray myotis were in alarming decline. This was especially true after the Tennessee Department of Health sponsored research to discover that a small proportion of gray myotis could contract rabies. Despite the fact that no human, or for that matter any other animal, is known to have contracted rabies from a gray myotis, warnings of this species being dangerous were broadly disseminated, and many colonies were destroyed out of the resultant fear. One of my primary gray myotis research caves was burned by its owners, despite the fact that the family had never been harmed over several generations. When health officials warned that their bats were dangerous, the family poured fuel oil into the cave and lit it in an attempt to exterminate the bats. Fortunately, these bats escaped through another entrance.

Unfortunately, uncounted thousands of other gray myotis were not so lucky. Paul Robertson, a fellow graduate student from the University of Kansas, joined me in the search for new gray myotis caves in July and August of 1968 and we experienced the wholesale destruction firsthand.

Indian Cave, on the Holston River northeast of Knoxville, was reported to shelter a large gray myotis nursery colony. Thus, it was one of the first caves we visited together. After it had been commercialized as a tour cave, its manager proudly explained how he and his staff had finally rid the cave of its bats. "We just poured kerosene beneath their roost. When we lit it we got 'em all. It was a mess! We had to haul wheelbarrows full of those stinking suckers out, but we've had no more problems." Of course, the only problem they ever had with bats was their own unfounded fear.

Paul and I easily located the former roosting room less than 300 feet inside. The stained ceiling where the bats had roosted indicated this had been one of Tennessee's largest remaining colonies. It likely had contained many thousands of bats. None remained – just the faint odor of burnt flesh and fur.

Next, we visited a cave where we found hundreds of gray myotis, both dead and dying. They had been attacked by vandals who had clubbed them with wooden sticks, left behind as mute evidence. Among the dead, we found 36 that I had banded. Then we ran into two graduate students participating in the Tennessee Department of Health search for rabies in gray myotis. They were using large hand nets to scoop hundreds at a time from nursery roosts, dumping mothers and young into large fertilizer bags to suffocate before being hauled off to a lab for testing.

In northern Florida, we searched for several caves near Marianna where colleagues had reported the state's largest gray myotis colonies. Of the two most important caves, one had been inadvertently buried beneath an expanding garbage dump, and the other, in the Florida Caverns State Park, had recently been opened to the public. There was a new parking lot in front and a sign inviting visitors to explore at their own risk.

As we entered, we had a pretty good idea what we'd find – sticks, stones, and bat skeletons. It was a sickening sight. The cause of death was clear. Unsupervised visitors had thought they were doing the world a favor by killing bats.

The bats had found this cave, with its multiple entrances and chimney-like air flow, ideal. When we complained to park managers, they seemed surprised that anyone cared about bats. They responded, "But this is the only cave in our park that doesn't have formations." Meaning, the only one that didn't require protection.

Gray myotis decline was so precipitous and seemed so nearly hopeless that Roger Barbour and Wayne Davis predicted the species' extinction in their 1969 book, *Bats of America*. They wrote,

"In the last few years human disturbance has threatened the very existence of this species," concluding that the gray myotis "probably faces extinction."

We were appalled by the massive, entirely needless devastation we were seeing. Nevertheless, over the years of my field research I had met many cavers and cave owners who completely changed their attitudes when I took time to explain the truth about bats.

White Bius, a farmer who owned an important gray myotis nursery cave about 40 miles north of Knoxville, Tennessee, illustrated how quickly and dramatically attitudes could change. When Paul and I asked permission to study his bats, he apparently welcomed us because he believed we would kill them. He said, "Fine, but please kill all the bats you can find." I suppose he thought it self-evident that scientists would understand why he didn't want bats on his property.

A deep stream flowed into and through much of the cave, probably explaining why his colony remained unharmed. Few people dared enter. Paul and I used a hand pump to inflate my one-man life raft, and I donned a sturdy pair of chest waders. Paul handed me a paddle, and I nervously allowed the current to pull me into the cave. I had no idea what lay ahead, but was determined to find out.

The warm stream, combined with the bats' body heat, raised the air temperature to 70 degrees at shoulder height. By clustering tightly together on a dome-shaped ceiling, the bats managed to raise their actual roost temperature likely 20 degrees higher, ideal for rearing young.

Soon I heard the high-pitched chittering of bats and looked up to see a colony of some 50,000 mothers and nearly full-grown pups off to one side and 20 feet above. Not wishing to disturb them needlessly, I quickly paddled back upstream, using my paddle to pry myself along one wall against the current. In so doing, I noticed the unmistakable, yellow-and-black-striped, outer wing covers of potato beetles discarded by bats that had caught them on the way home.

Having observed that White Bius was growing potatoes nearby, I carefully grabbed a sample. Later, when I showed him the conspicuous wings and asked if he knew what insects they came from, he immediately recognized them as one of his worst crop pests. The expression on his face was easy to read. "Them bats eat potato bugs? How many do they eat?"

I answered, "It depends on which kinds of insects they find on a given night. When they're mosquito-sized, just one bat may catch a thousand in an hour. All combined your colony may eat up to a hundred pounds of insects in a single night, including your potato beetles." His response, "That's a lot o' bugs!"

Nothing more needed to be said to turn his attitude around 180 degrees. A year later when I returned, Mr. Bius had become an ardent bat protector. He proudly informed me, "I've been a-thinking, my bats is worth more than five dollars apiece." His protected cave became one of my favorite research sites, and his quick and lasting conversion was one of many that helped give me courage to found Bat Conservation International more than a decade later.

My petition for the U.S. Fish and Wildlife Service to list the species as federally endangered was granted on April 28, 1976. Initial progress was frustratingly slow. Bats in general still ranked right between rattlesnakes and cockroaches in public popularity. In fact, even my closest friends and colleagues thought I was crazy when, in 1982, I announced that I was resigning from a wonderful research position at the Milwaukee Public Museum to found an organization devoted entirely to the conservation of bats.

Nevertheless, the truth about bats proved far more powerful than even I had anticipated. I would go on to better document bat values and needs, learn photography, and develop my ability to entertain both individuals and audiences with fascinating bat facts. In doing so, I'd meet some of the finest people and organizations ever. Time after time, key individuals joined Bat Conservation International, while The Nature Conservancy, the American Cave Conservation Association, the Fish and Wildlife Service, the Tennessee Valley Authority, and many other private and governmental organizations pitched in as partners. Together, we would exceed even my most daring conservation dreams.

For the gray myotis, the Sauta Cave and Fern Cave National Wildlife Refuges were created. And over the next 25 years, Pearson and nearly all of the gray myotis's most important roosting caves throughout its range would gain long-term protection. Results have been dramatic. The Pearson population more than doubled. And the one in Hubbard's Cave, one of my study sites in central Tennessee, recovered dramatically. Its population, which was in precipitous decline at the time of acquisition by the Tennessee Nature Conservancy, quadrupled in size to more than half a million. Bellamy Cave's population in northwest Tennessee grew from just 65 individuals to more than 150,000 after I convinced the owner to protect it from vandals.

These results are extremely encouraging. Today, there are millions more gray myotis than when their extinction was predicted.

▶ Merlin climbs high above the cave floor searching for banded bats in Fern Cave, Alabama.

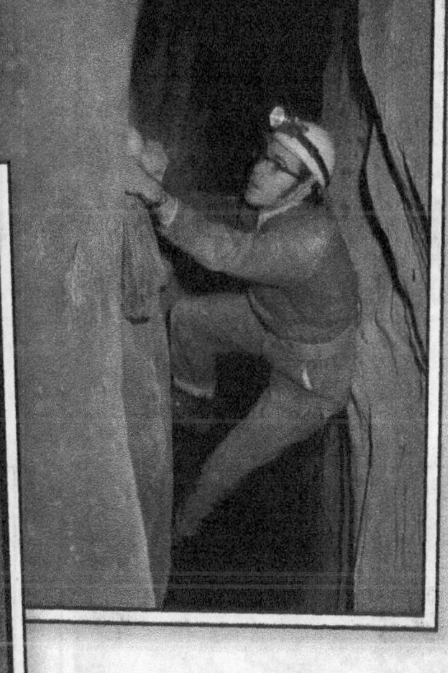

◀ The author crawls into a hollow tree in search of frog-eating bats in Panama.

▲ Befriending epauletted bats photographed for *National Geographic* in Kenya.

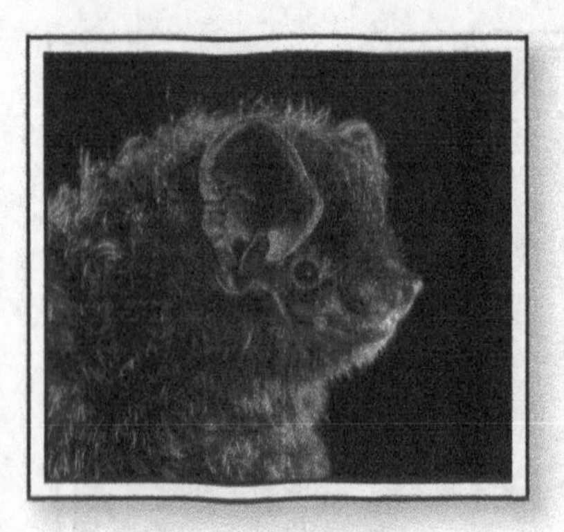

Silver-haired Bat
North America

Honduran White Bat
Central America

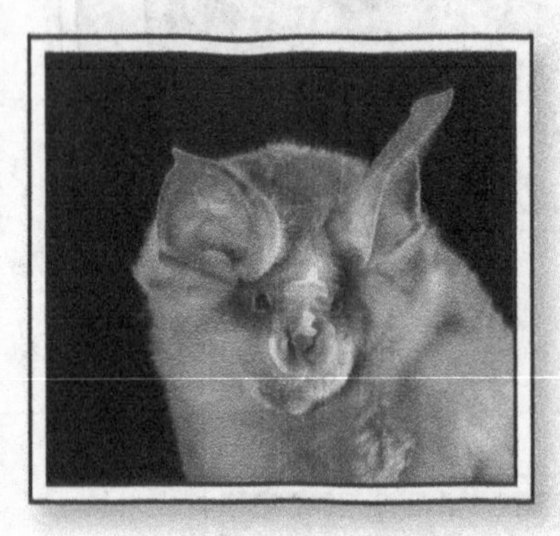

Mehely's Horseshoe Bat
Eastern Europe to Middle East

Hairy-legged Vampire Bat
Latin America

Hoary Bat
North and South America

Eastern Red Bat
Eastern North America

Mexican Long-tongued Bat
SW United States to Honduras

Greater Short-nosed Fruit Bat
South & Southeast Asia

Chapin's Free-tailed Bat
Central & Southern Africa

Gray Myotis
Southeastern United States

California Leaf-nosed Bat
Southwestern U.S. & Northern Mexico

Madagascan Fruit Bat
Madagascar

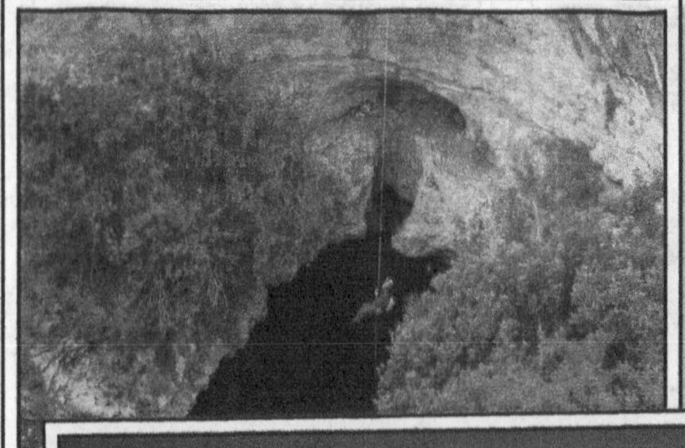

◀ Merlin ropes into a Brazilian free-tailed bat cave in West Texas.

▲ Millions of tourists have safely observed bats close-up in Austin, Texas, without harm, and bat watchers add approximately 12 million dollars to Austin's economy each summer.

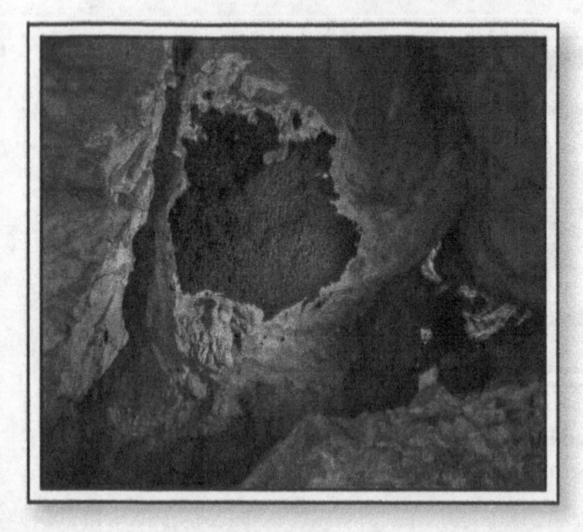

◀ A cluster of 100,000 gray bats hibernates at 32 degrees in Pearson Cave, Tennessee. Gray bats live in caves year-round, migrating between exceptionally cool ones for winter hibernation and warmer ones for rearing young in summer.

◄ A common vampire bat feeds on a sleeping cow in Costa Rica. This species has overpopulated thanks to the introduction of livestock, and now must be controlled with great care to avoid unintended harm to beneficial bats. Vampire bats live only in Latin America.

▲ An epauletted bat takes a fig in Kenya. Fruit-eating bats are the most important long-distance seed dispersers of wild figs worldwide.

▲ Young epauletted bats appear to learn the scent of edible fruit from licking their mother's lips.

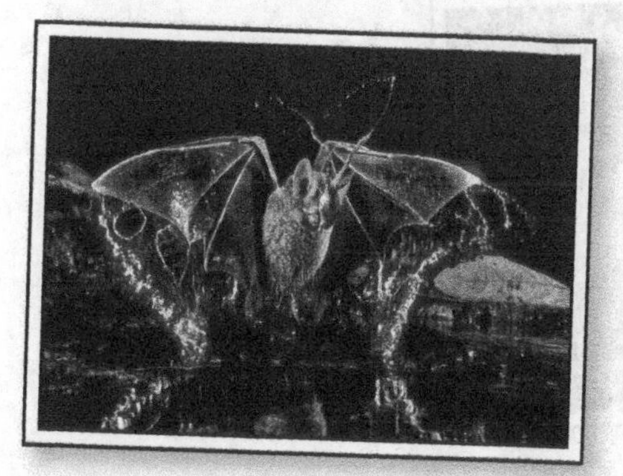

◄ A frog-eating bat snatches a courting túngara frog from a rainforest pond in Panama. These bats can identify frogs by their mating calls and remember even briefly learned new calls for several years.

◀ Seldom-seen greater bonneted bats live high up on cliff faces across much of western North and South America.

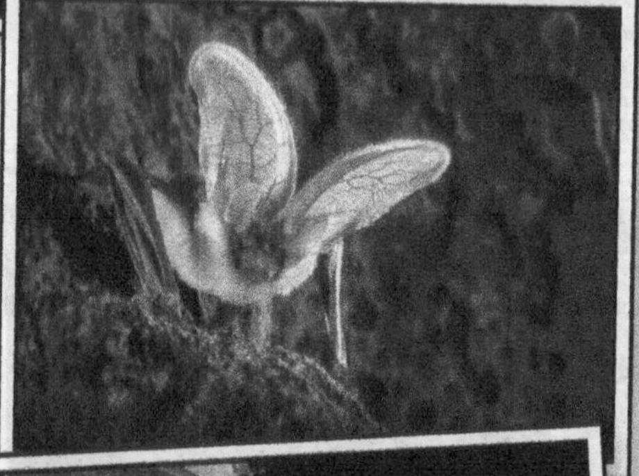

◀ Spectacular spotted bats are widespread in western North America but rarely seen. Their extra low echolocation calls are audible to humans but not to the moths upon which they feed.

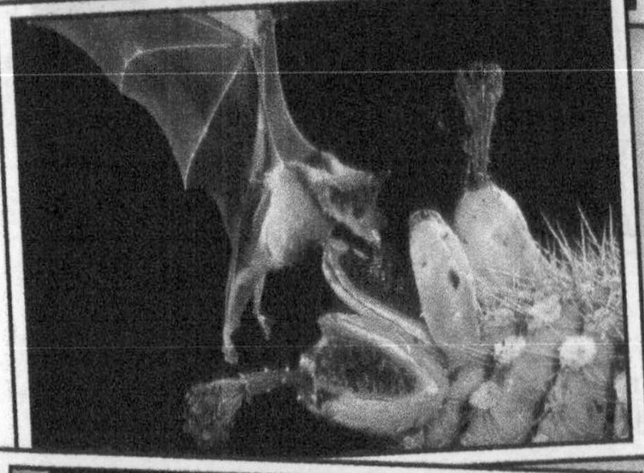

◀ Long-nosed bats are extremely efficient long-distance seed dispersers for several cactus species in deserts of the southwestern United States and Latin America. This one is feeding on saguaro fruit.

◀ Giant cacti, from this cardon to organ pipe and saguaro, provide key habitat for many animals, but in turn rely heavily on long-nosed bats for pollination and seed dispersal.

◀ Mother free-tailed bats bear just one pup per year but can identify their own from among millions of others, something scientists once thought to be impossible.

▶ In just one night a free-tailed bat can eat 20 to 40 migrating moths that could each lay 500 to 1,000 eggs on Texas crops. The combined impact of millions of these bats significantly reduces the need for costly pesticides.

▲ Ten to twenty million free-tailed bats cover the walls of Bracken Cave in Texas at approximately 200 adults per square foot. Their impact in keeping insect populations in balance is enormous.

◀ Millions of wrinkle-lipped bats, emerging from their protected roost in Khao Chong Pran Cave in Thailand, protect nearby rice crops by consuming tons of planthoppers nightly.

▶ Durian fruits constitute a multibillion-dollar industry in Southeast Asia, but require bats to pollinate their flowers, in this case a cave nectar bat from Khao Chong Pran Cave in Thailand.

◀ Wild banana plants, the source of all commercial varieties, require bats like this short-nosed fruit bat in Thailand for pollination. These plants are invaluable sources of genetic material needed to improve cultivated varieties.

CHAPTER 3
TRACKING BAT NIGHTLIFE

DURING MY GRADUATE RESEARCH, I had learned much about gray myotis migration and roosting needs, and that they fed predominantly over water. But they obviously preferred some locations over others. Once these bats were recognized as endangered, federal and state management agencies needed to know more about their feeding requirements.

Funded under a contract with the Tennessee Valley Authority (TVA), my former wife, Diane Stevenson, and I set out to discover where these bats went to feed and what they were looking for. We decided to follow gray myotis from a nursery colony of 8,000 that lived in Oaks Cave, in the Chuck Swan Wildlife Management Area, 25 miles north of Knoxville, Tennessee. I had worked there during my graduate research and knew the area well.

The cave opened into a cool sinkhole at the base of a moss- and fern-covered limestone cliff, shaded by towering beech, hickory, and maple trees. It was bordered on three sides by the 34,000-acre Norris Reservoir, one of America's cleanest lakes, with numerous islands and over 800 miles of shoreline, ideal for studying the feeding behavior of bats that hunt over water.

We arrived for our first summer's work in May 1976, proud as could be of our new Volkswagen Campmobile, which we had specially outfitted for bat studies. Brackets on the rear conveniently held two bat traps. Labeled drawers housed our equipment, bat bands, maps, and data records. The pop-up top allowed ample sleeping space, and we even had a small kitchen.

The next morning we met with Ralph Jordan Jr., TVA's environmental scientist, at his office in Norris to pick up our 12-foot Boston Whaler and 40-horsepower outboard motor as well as a couple of night-vision scopes, all on loan for the summer. After years of being an impoverished graduate student, always having to make do with minimal equipment, having all these researcher "toys" and actually getting paid to chase bats in a beautiful setting was exciting, though we would face unexpected challenges – some nearly fatal.

Our first goal was to follow feeding gray myotis by boat to see where they were going. Since most of the reservoir was navigable, at least in theory, we should be able to follow them using a powerful spotlight. We had developed a system for using reflector-tape-covered bands so we would know their sex and whether they were adults or juveniles. A red band on a right wing indicated an adult female, while the same color on the left would identify the wearer as an adult male. Juveniles would be banded a month later, as they learned to fly, and they would get white reflector tape, again females right, males left.

That night we set a bat trap at the cave entrance and quickly captured 400 adults, mostly nursing mothers. We slipped bands over forearms, then partially closed them so that they resembled bracelets. As we released banded individuals to fly off toward the reservoir, spotlight reflections from their bands made them glow like fireflies.

We could hardly wait to follow them by boat. What kinds of insects were they looking for, and where would they go to find them?

As the sun set the following evening, we waited expectantly in our Boston Whaler, motor idling, ready for our first chase, armed with our spotlight, topographic maps, a compass, an extra fuel tank, and warm coats. A cold front had passed through during the day, triggering thunderstorms and rain. We now had a clear sky, but the temperature was a chilly 46 degrees.

To reach the reservoir, our bats had to fly through a third of a mile of forest, so we couldn't be exactly sure where they would emerge. We listened with a hand-held bat detector, a device that detects bats by converting their ultrasonic echolocation signals to audible frequencies that can be heard by humans. Each time we heard the staccato beeping sound of an approaching bat, we briefly hit it with our spotlight to see if it was one of our marked individuals.

Red bats (*Lasiurus borealis*) were the earliest fliers, but they stayed high above the reservoir. When a steady stream of bats began swooping low over the water, we knew they were gray myotis. Based on our estimation of colony size, only about 1 in 20 should be marked.

Just before nine o'clock we spotted the first bright red reflection. It was a female, and we followed about 75 feet behind her, me standing, steering wheel in my left hand, spotlight in my right. I kept her just well enough illuminated to see her reflection without frightening her, while Diane timed her movements with a stopwatch and plotted our course on a topographic map. We were able to keep up with her for close to a mile, at which point she cut across a shallow area where our boat couldn't safely go.

Field notes for an individual we followed on July 2 read, "RR

[Red-Right, meaning adult female] was followed by boat, using edge of spotlight beam, to Clinch River entrance past Point 19, a distance of 3.2 miles in 16 minutes (9:30–9:46 PM) for an average speed of 12 miles/hour; she foraged most of the way, making approximately one capture attempt every three seconds, taking most prey from water surface. Moths ½–1 inch long seem to be preferred, though they amounted to less than one percent of captures. She slowed and foraged more when passing through dense swarms of hatching insects which often covered 100 yards or more in diameter."

We sometimes followed several of our marked bats in quick succession, but seldom for as far as we had hoped. One evening we had tailed a feeding bat for a couple of miles, obtaining excellent data, when she flew between a barge loaded with partiers and the shore about 100 feet away.

I hated to disrupt the party with our wake, but was determined not to lose our bat, so I steeled myself for a possibly angry outcry. Suddenly screams from the party sounded far more desperate than angry, and Diane and I hit the deck just in time to avoid decapitation by a quarter-inch-diameter steel cable they had strung to a tree on shore. The bat all but forgotten, we circled back with profuse thanks for their timely warning. They were fascinated to hear we'd been chasing a bat, and we spent the next 30 minutes answering questions.

"Aren't they all rabid? Aren't you afraid of them? Why would you want to study them?" Their questions were typical of what bat researchers had to deal with in the mid-1970s, and I'd like to think that we replaced their misguided notions with a better understanding of bats' gentle and harmless nature.

Another evening we were far from our base camp when a heavy fog blanketed the lake. We struck a rock that destroyed our shear pin, making our outboard motor inoperable. At 2:00 AM there wasn't a lot of help to be found, but we could hear blaring music in the distance. Using a canoe paddle, we inched toward it only to discover a drunken party of folks best described as redneck hippies. They had long hair and drugs, and a bad reputation: TVA's Ralph Jordan had warned us to give them a wide berth. They reportedly had shot at TVA boats.

Fearing the worst, we paddled up to their shoreline party and explained our desperate need for a shear pin to fix our motor. A couple of rather heavy dudes waded out to us, looked at our motor, and yelled to someone onshore. Minutes later they had installed a new pin, and we were most gratefully on our way. Another lesson learned; we never went anywhere again without spare shear pins.

We gradually got better at avoiding trouble, as well as at tracking our bats. We even learned the best coves in which to seek shelter during brief but sometimes intense thunderstorms that could surprise us as far as 25 miles from base camp. Quite aside from risking collisions with cables and barely submerged rocks and logs in the heat of a chase, we had to be constantly on the lookout for thunderstorms, some of which could be quite a challenge for people in small boats.

One night in May, Diane and I decided to check for gray myotis from Oaks Cave on the far side of the peninsula on which the cave was located. We'd seen hundreds of our bats heading into the forest in the opposite direction from the nearest lakeshore and needed to know if they were feeding in the forest or over the reservoir on the far side of it.

We took a compass bearing on the direction they were headed, then calculated where they should emerge if indeed they were planning to feed over water. Their trip over land was only 6.5 miles, but our boat trip around the peninsula to meet them required 25 miles. We left early and waited for them at dusk, again listening for their echolocation calls with a directional bat detector.

We could only hope they had held to their original compass heading. Otherwise, we could be in the wrong location to successfully intercept them. We didn't have long to wait before gray myotis began to emerge from the forest, and a few minutes later we spotted our first marked individual. In unison we breathed a big sigh of relief. "They're ours!"

Nevertheless, within another half-hour, we began to see flashes of lightning on the far horizon. Knowing we shouldn't risk getting caught so far from our base camp, Diane nervously cautioned, "Did you see that? I suspect it's headed this way." But in my enthusiasm for following these new bats, I assumed there was a good chance it would miss us.

I was wrong, and with unbelievable rapidity we found ourselves in for a direct hit, still some 20 miles from our camp. By this time we had learned the importance of finding a sheltered lagoon. Pelted with a torrent of blinding rain and with lightning and growing whitecaps all around, we raced for the nearest shelter, finding a lagoon barely in time to avoid being swamped. We pulled into a narrow spot and huddled miserably under our ponchos. The storm abated within an hour, and we were able to resume our return journey, chilled to the bone by the time we got back in the wee hours.

A few days later, when really severe weather was forecast, we opted to take a night off, remaining with our vehicle at Oaks Cave. As the storm front approached at sundown, we abandoned our vehicle and took shelter in the cave entrance. Within minutes, even the largest trees were swirling around like saplings, and the lightning was so constant that we didn't need our headlamps to see.

Moments later, we were amazed to see thousands of our mother gray myotis emerging. They must have been truly desperate to feed their pups, flying directly into what looked like a giant, flickering light bulb amid a now continuous roar of sound. I often wondered what happened to them.

While seated on an entrance rock with Diane, I couldn't resist recalling a time years before when, as a graduate student working on my Ph.D. research, I had visited this cave with volunteer Dave Weaver. Generally brave enough to face nearly anything, he had balked at having to crawl through loose slab rocks to enter. He had asserted, "That's a rattlesnake den if I ever saw one." I had assured him I'd never even seen so much as a garter snake there. Reluctantly, he'd finally followed me in.

Later that night, as we exited, I thought I heard a pack rat scamper away. Then I heard a second similar sound, and as far as I could tell, it stopped in a small clump of ferns just in front of me. Always fascinated by all wildlife, I said, "Dave, there are pack rats here. I believe one is hiding here in these ferns. Watch, and I'll nudge it out."

With that, I slowly moved my toe beneath the fronds, and, to my astonishment, out crawled a nearly three-foot-long copperhead. Dave unhappily responded, "I see why you've never seen snakes here. You simply can't tell them from pack rats!"

I figured this wasn't a good time to tell that story to Diane. The next morning we found that a quarter-mile-long swath of forest had been demolished as if the trees were mere matchsticks. A tornado had passed within 100 yards of Oaks Cave.

By the end of our first summer, we still hadn't seen another copperhead, but that wasn't our last threat from fierce storms. In total, we had spent 93 hours following our bats in all kinds of weather. We'd tracked our Oaks Cave bats along more than 225 miles of lake shoreline, sometimes covering more than 60 miles in a single night. We had learned where our bats went to feed as well as their travel routes and had marked these on topo maps. After emerging from their nursery cave at dusk, they would feed almost anywhere over the reservoir as they traveled at leisurely speeds to more distant locations where small groups appeared to be guarding specific feeding territories later in the night. Many of these sites were 6 or more miles away, and one was nearly 15 miles. Why travel so far with so much lakeshore in between?

To answer that question, we would need to simultaneously compare used versus unused sites to see how they differed. This would require at least one more summer and another person on our team, so we were delighted to receive a letter from University of Tennessee graduate student Alan Rabinowitz, inquiring about the possibility of conducting thesis research on gray myotis under my guidance. I called him and invited him to work with us that summer, studying gray myotis feeding behavior on the Norris Reservoir, but I also warned him that we'd often be working all night, with minimal time for daytime naps, and that conditions might be tough at times. Having no idea what he was in for, he assured me it was no problem and that he could handle it. Of course, at the time I had no idea that our very lives would be challenged. Alan worked out regularly and was in excellent athletic condition. Certainly he could handle a little sleep deprivation.

Our first "get acquainted" field trip was planned for early April 1977. My plan was to have an easy motorboat trip across the Norris Reservoir, culminating at Oaks Cave. We would follow the bats to feeding sites and observe their early spring behavior. Not in our plans was the arrival of a powerful cold front with flooding rains and high winds, followed by subfreezing temperatures.

When we arrived at a dock in a sheltered lagoon, we were surprised to learn that the only available boat was a relatively unstable jon boat. Nevertheless, we felt confident as we loaded expensive night-vision equipment and prepared to head out across the three-quarters-of-a-mile-wide reservoir. I even welcomed the exceptional cold, wondering how the bats would respond.

Then, just as we approached mid-lake, we were hit by high winds and towering waves. I yelled "Grab the equipment!" as our boat suddenly nosed under a five-foot-tall whitecap.

I wanted to record gray myotis feeding behavior on an unusually cold evening, but we hadn't bargained for this. As our boat filled with water, whipped by powerful gusts of wind with just-above-freezing temperatures, Alan later admitted that his only thought was, "We're going to die." Having just acquired $7,000 worth of new equipment, I was totally focused on not losing it.

Fortunately, everything was in watertight containers, and we were able to hang on to it and to the swamped boat as we were gradually blown toward shore. First we feared we'd never make it. Then, when we did, we realized we were soaking wet, freezing cold, and far from help with no hope of building a fire. We had run ashore in a wildlife management area miles from the nearest help.

Badly shaken, we knew our jon boat was our only hope, though it was woefully inadequate. It was only 12 feet long and flat-bottomed. We didn't even expect the motor to start again, but in desperation we emptied the boat and tried anyway. To our amazement, after several tries it sputtered to life.

From a shallow cove, we were able to relaunch with great trepidation as we again approached the angry waves. Al looked at me as if I were stark raving mad. But I responded, "Our only hope is to reach the shelter of the cave. We're within a half-mile." We would soon be hypothermic if we didn't find shelter.

I had learned one key lesson: Don't try to steer a jon boat straight into towering whitecaps. We were greatly relieved to find that by angling up and down the waves, instead of meeting them head-on, we could stay afloat and gradually move ahead—just barely!

By the time we arrived at the cave, we were borderline hypothermic, but I figured that, while there, we might as well set up the equipment and record some observations of the bats, which we were amazed to see emerging despite the storm. I suspect that Al would have mutinied except that the prospect of re-crossing the lake to return to the boat dock was more frightening than freezing to death where we were. In fact, the return trip to the dock would be three and a half miles.

Hoping the wind would die down, we waited until nearly nine o'clock before attempting the return. We had a powerful spotlight, but it didn't make the waves any less intimidating. We seemed to be almost constantly airborne, skipping from one whitecap to the next. By the time we reached the dock, Al, despite his strong athletic condition, had to ask for help getting out of the boat. He said, "I'm not sure whether I'm just hypothermic or too terrified to stand up!" Amazingly, he decided to go ahead and work with me on his thesis. Years later, he wrote to say, "I knew after surviving with you that I could do anything."

Nearly drowning was only the first of several challenges Alan would face that season. He had grown up in Brooklyn, New York, isolated from outdoor experiences, so one night when he had to go to the bathroom in the middle of nowhere he made the mistake of wiping with a handful of poison ivy leaves. Needless to say, he walked with an interesting waddle for the next week despite a trip to see a doctor. Then one morning, a sticky trap for insects fell onto his head, requiring seemingly endless scrubbing with kerosene to free him. It was a four-inch-diameter plastic conduit coated in a product called Tanglefoot, one of the most obnoxiously sticky, difficult-to-remove substances I've ever encountered.

Our primary research goal, in collaboration with Alan, was to document insect availability at sites that were used versus unused for late-night feeding. We trapped simultaneously at pairs of locations at half-hour intervals all night long to compare insect species and sizes present and the times of their availability.

The results were striking. Used versus unused sites differed very little during the early evening period of massive insect hatches, but territories guarded by bats later at night consistently provided late mayfly hatches. Over an entire night, some unused sites actually out-produced guarded sites in total numbers of similar-sized insects. The critical difference was in the timing of insect availability.

Our insect traps revealed that 90 percent of an entire night's flying insects were available only in the first hour and a half after sundown, leaving just 10 percent spread over the remainder of the night. During the early evening period of major aquatic insect hatches, groups of up to 50 gray myotis would spread out over large areas of lake, converging to feast when hatches were discovered. Later, small groups would join forces to tenaciously guard territories along steep, rocky shores that favored late-night mayfly hatches. These were in short supply, explaining the need to travel long distances to find them.

Insects were clearly relying on what we called a "feast or famine" strategy. They emerged in such huge hatches at dusk, many kinds all at once, that their bat predators couldn't possibly catch them all. Then they'd almost magically disappear before bats could find enough additional food to sustain their remaining energy needs.

From my prior studies I knew that the Oaks Cave nursery colony of gray myotis was rearing young in an unusually cool roost where energy demands to stay warm were high, probably requiring multiple meals per night to support the pups.

Close-up photos, taken with a camera and flash mounted atop my night-vision scope, documented that feeding territories were owned exclusively by nursing mothers, who were easily identified by their milk-laden breasts. Territory owners detected potential intruders from at least 20 feet and chased them away. Most were merely warned to leave, but those who persisted were sometimes actually struck and knocked into the lake.

At our best-studied territory, we were able to capture and band the most dominant female from a group of six nursing mothers, placing a reflective band on her arm. Other bats were quickly driven off, typically by the dominant owner, sometimes with varied amounts of help from the others who shared the site. Since males and females without young require only about half as much energy as nursing moms, they apparently don't need to guard territories.

When hatches were exceptionally poor, our dominant female often drove even her group-mates to the periphery. Conversely, on exceptionally good nights, when hatches were large, feeding territories were not defended, and everybody simply shared.

We surmised that dominance in the pecking order probably increased a mother's likelihood of producing surviving offspring. This may explain why yearling female gray myotis seldom rear young despite mating in the previous fall. Until a territory owner dies, or becomes too old to successfully defend her position, younger females may not be able to obtain the energy required to rear a pup. Subsequent observations would document that our marked mother continued to dominate her territory for at least four more years.

Unlike most other mammals, bats appear to have extraordinary mechanisms for population regulation. When energy needs cannot be met by a certain point in pregnancy, a mother may simply resorb her embryo.

We now know through the meticulous work of United Kingdom bat biologist Roger Ransome that some bat mothers are what he calls "super moms." By tracking the annual reproductive success of marked greater horseshoe bats (*Rhinolophus ferrumequinum*) over several decades, he has documented that some mothers are dramatically more successful than others at producing surviving offspring.

I suspect that our code-marked mother, who dominated all others, may have been a super mom. We often observed her and her "friends" all night long. Her territory was curved around a rocky point and was 60 feet wide by 300 feet long. The site would be patiently patrolled for an hour or two at a time, even when no new hatches occurred.

While observing this territory, we couldn't help noticing that a similar group of bats was simultaneously guarding a territory directly above our gray myotis. While our bats rarely chased an insect more than 15 feet above the water, this second group of bats routinely fed from 15 to 40 feet up along a rocky bluff. Neither group seemed to pay much attention to the other.

Who were these bats? Were they male gray myotis or were they another species? To find out, Diane and I spent hours one hot afternoon rigging a 42-foot mist net between two trees over the bluff. We knew that small, insect-eating bats would be difficult to catch in a not-too-well-set net, but we were hopeful. At sundown we waited patiently.

Shortly there was a *whoosh, whoosh, whoosh* sound from behind and, too late, we realized it was a great blue heron. Needless to say, catching such a large bird, with its long, sharp bill for a weapon, is a bat man's nightmare. Its weight sagged our net nearly to the ground, but we were totally unprepared for extricating an enraged heron. As I rushed to the net, Diane yelled, "Watch out for its bill!"

Very carefully I got ahold of its neck and yelled, "Now grab its feet!" While the heron flapped its long wings, we had to tear it from the net. With the net destroyed, we had to give up for the evening.

The next night we were back, this time armed with fly-fishing tackle. I've long been a catch-and-release fly-fisherman as a hobby and was aware that trout fishermen, fishing at dusk, frequently catch feeding bats by accident. My idea was to climb to the top of the bluff equipped with a fly rod, an exceedingly fine leader and tippet, and a tiny trout fly with a debarbed hook.

As Diane watched nearby, I allowed a gentle breeze to waft my fly into the bats' feeding zone. Almost immediately one of the bats caught my fly, but I had as yet no experience in how to land a flying bat. It headed for the nearest tree as I attempted to take up slack and pull it into reach. I slipped on loose pebbles, inadvertently releasing more line. As I struggled to get up, I was yelling at Diane, "See if you can get ahold of the line. Keep it out of the trees!" Though it seemed to take forever, she soon had it in her gloved hand.

It was a nursing mother tri-colored bat (*Perimyotis subflavus*), completely unharmed by our novel capture technique. Over the next ten minutes we caught several more, all mother tri-colored bats. In our excitement we only later noticed bass fishermen watching in amazement from their nearby boat. Promptly releasing the bats, we could only imagine what the fishermen must have been thinking. Since the bats would have been too small for them to see, there must have been some whopper stories circulating among fishermen that summer.

On another night we used a small drop of surgical glue to attach a tiny, hand-blown glass sphere, filled with activated Cyalume (a chemical light), to a bat's back in order to observe its feeding behavior. In those days, bat researchers were attempting to use chemical lights to follow bats. The glowing spheres soon fell off with little inconvenience to bats, and they turned out not to be very effective for research.

Not surprisingly, our bat didn't behave normally with a glowing light on its back, but minutes later we heard exclamations from fishermen anchored nearby. "My God! What in the world is that?"

"It's too big for a firefly."

"It's a bat!"

"You're crazy. Bats don't glow."

Our findings were incorporated into the U.S. Fish and Wildlife Service's Endangered Species Recovery Plan for the gray myotis, and Alan went on to become a leading conservationist, referred to by *Time* magazine as "the Indiana Jones of wildlife conservation." He became an expert on large cats, from jaguars to tigers. During his first major project he survived a plane crash in the jungle and lost his field assistant to the bite of a deadly fer-de-lance snake.

CHAPTER 4

INVESTIGATING VAMPIRE BATS

"OUCH!" It was my first time to be bitten by a common vampire bat (*Desmodus rotundus*). Soon, I was surrounded by curious Chamula Indians who weren't favorably impressed with my intelligence when they saw my profusely bleeding finger. Judging from their expressions, they were thinking something along the lines of, "Only a fool plays with vampires."

During the summer between my sophomore and junior years in college, I had convinced my family to buy a Jeep and join me in collecting small mammals for the American Museum of Natural History in New York. This was my first professional job, and I had set a 42-foot-long mist net in the entrance to a cave along a mountain ridge near Pueblo Nuevo in Chiapas, Mexico.

I'd been using a leather glove on my left hand, while extricating the bat from the fine-threaded net with my ungloved right hand. And since this was my first vampire, I was unaware of just how clever they could be in defending themselves. Now I was missing a pencil eraser–sized chunk of skin from my index finger, and I was bleeding profusely thanks to an anticoagulant in the bat's saliva.

The cave's large sinkhole entrance, though located in lush tropical vegetation, was easily visible from the road above, hence the gathering of curious Chamulas, mostly men wearing dark-colored wool ponchos. These people spoke the Tzotzil language, in which they called themselves the *Sotz'leb*. Translated into English, this means "bat people," perhaps explaining their particular curiosity about a bat man.

When the vampires had left to feed on local cattle, they had seen my net and skillfully avoided it. Nevertheless, when they attempted to return an hour later, I caught them easily. They were so heavily engorged with blood that they appeared on the brink of bearing twins and, like near-term humans, they were no longer agile. In just moments, I caught more than 20.

Rapidly disentangling them from my net, as more plummeted in, was a young bat enthusiast's nightmare. In a hurry, and distracted by my unexpected visitors, it was virtually inevitable that I would get bitten.

I had strung my bat net between two poles like a volleyball net. The nylon netting was as fine as hairnets worn by servers in cafeterias. This material was held in place by five horizontal main strand cords. Six-inch loops at the ends of each main strand were slipped over the tops of two poles. By moving the loops up or down, I could adjust the main strands to form horizontal pockets of netting that hung below.

Bats striking the net fell into one of these pockets unharmed but became too entangled to escape. Removing a vampire from such a fine-threaded net takes a bit of finesse.

Vampire bats, despite my first unfortunate experience, are actually gentle when not defending themselves against what must initially be perceived as an attack by a giant spider. Though much maligned for their blood-feeding habits, vampires are actually quite sophisticated. They have a social order similar to that of primates. Like humans, they share food and information, adopt orphans, and practice reciprocal altruism. That is, they are most generous to those who have helped them in the past, a trait that is rare beyond humans, chimpanzees, and wild dogs.

Scientists who study vampire bats in captivity find them to be extremely intelligent, even affectionate. Nutritional physiologist Claudia Coen reported that whenever she removed and returned a member of her captive colony of white-winged vampires (*Diaemus youngi*) to their cage, "They would greet each other, both vocally and through a ritual of pressing their noses into each other's armpits and wrapping their wing membranes tight around in a 'hug' of welcome." They were shy of strangers but greeted her as one of their own. She said, "They'll nuzzle their noses between my fingers in greeting. They're enchanting!" Coen also reported that her bats liked to cuddle, and they played with exuberance.

In 1990, Dieter Plage worked with me at the La Selva Biological Station in Costa Rica to film a common vampire feeding on a chicken. We needed a tame one that would behave naturally in front of a noisy movie camera, so after capturing one at a local roost, we gave it to our field assistant, Bert Grantges, a bat-loving boy of 13. Bert was so excited about getting to see and handle a vampire bat that he could hardly contain his enthusiasm. I instructed him, "Handle it gently with gloves till you're sure it is no longer feeling threatened." He took the bat to his room, and just two hours later, I heard him calling, "Merlin, come look." The vampire was peeking out from Bert's shirt pocket as he exclaimed, "It's as tame as a hamster!"

The only problem for the shoot was finding a chicken that

would stay asleep during our filming. Twice, we brought a hen that refused to sleep on a perch in front of a noisy camera and movie lights. Bert's vampire, however, performed flawlessly. When we finally found a sleepy bird, the bat cautiously stalked it, moving a few inches, pausing, then advancing again. When the vampire nuzzled and began licking a toe, the hen appeared to relax, perhaps mistaking it for a chick. That vampire is now immortalized in the documentary *The Secret World of Bats*. Bert hated to release his newfound friend.

Vampires typically live in small groups consisting of several mothers and their young, guarded by an adult male. Like humans, they rely on a greatly extended period of juvenile learning. Common vampire mothers nurse their young for up to nine months, at least six months longer than any other similar-sized animal.

Feeding on blood is risky and requires extensive training. Because blood contains negligible fat, which can be stored as an energy source for later use, vampires who miss just two meals in a row normally starve if not helped by a food-sharing roost mate. Young hunters must learn the habits of prey, how to detect sleeping animals, how to find the right location for a rewarding incision, how to open a wound and maintain blood flow without awakening the victim, and how to make a quick escape in case something goes wrong.

Even after successfully stalking a sleeping animal and making an optimal incision, vampires must prevent blood vessels from closing or the blood from clotting for the duration of an approximately 20-minute meal. And, since blood is mostly water, vampire kidneys must be extremely efficient, allowing them to concentrate protein by urinating almost as fast as they drink. Finally, in order to obtain sufficient nutrition, vampires sometimes must drink up to 60 percent of their body weight before flying home. That's the equivalent of a 150-pound human eating 90 pounds of food in one meal!

These bats bear little resemblance to the vampires of popular legend. Mythological vampires are humans who return from the dead to feed on the blood of the living. Vampire stories originated in Asia, Africa, and Europe, where blood-feeding bats have never lived, and they preceded the discovery of blood-feeding bats in Latin America by thousands of years. The term *vampire,* of European origin, was quickly applied to blood-feeding bats, once stories were brought back by Columbus's crew.

Though awareness of vampire bats is now nearly universal, in reality only three kinds of bats, far less than 1 percent of the world's bat species, feed on blood. These live only in Latin America, and only one of them, the common vampire, causes problems for humans. It prefers the blood of mammals and has overpopulated where forests have been replaced by cattle ranching. In such areas it periodically becomes a costly nuisance by spreading disease among livestock during rabies outbreaks. It also may bite poor people who sleep without protective mosquito nets or screened windows. White-winged and hairy-legged (*Diphylla ecaudata*) vampires are relatively rare and feed only on birds.

The common vampire is seldom common except where livestock have been introduced. After I graduated from college in 1964, Charles Handley hired me to codirect his Mammals of the Smithsonian Venezuelan Project. For the next couple of years, my field teams and I systematically sampled many thousands of small mammals, and we rarely encountered a common vampire far from humans who kept pigs, cattle, horses, or poultry. We also seldom found them near villages of indigenous people who didn't keep domestic animals. In sharp contrast, in ranching areas, common vampires often ranked among the most abundant mammals.

At one location in southern Venezuela, local people suggested I contact the government's vampire control men to help us find bats. When I went to their office, they were delighted to meet me and offered to take me with them on a vampire control trip. They were very proud of their latest equipment, a military-style flame thrower.

Several miles down a dusty, dirt road, we arrived at the first "vampire" roost, a concrete culvert about six feet tall, beneath the road. As we stopped the pickup and got out, one of the men quickly hoisted the flamethrower onto his back. It was fairly simple. A couple of fuel tanks in a special backpack were held in place with shoulder straps. A hose, pistol grip–type trigger, and nozzle were hand-held in front. When the trigger was pulled, a stream of gas would be ignited and sprayed up to 30 feet, incinerating anything in its path. I made sure I was in the rear.

We carefully climbed down through roadside vegetation until we reached the culvert entrance. Then, in Spanish, I asked, "Could I please see the vampires before you kill them?" They answered, "*Sí, como no?*" Yes, why not? As I shined my electric headlamp into the dimly lit tunnel, I wasn't surprised when I didn't see a single vampire. Several hundred fruit and nectar-eating bats had been about to be cremated. My hosts were quite surprised when I explained that their presumed "vampiros" were instead highly beneficial species that helped plant trees and pollinate some of their most valued plants.

There were vampire bats in the area. We could see the blood-stained necks of nearby cattle bitten the night before. What my hosts didn't realize was that common vampires form small, inconspicuous colonies, even living in people's open-topped wells.

Next, we drove to a cave deep in the forest. I could imagine their line of reasoning: if there were thousands of bats living in a cave, they must be vampires. After a short walk through the relatively cooler but humid rainforest, we came to a limestone bluff in which we could see a gaping cave entrance. Again, I asked for a first look, and again, none of the bats were vampires, just thousands of small insect-eating bats of at least three species.

In a second cave, we found a group of about a dozen common vampires. They were roosting in an inconspicuous side passage well beyond a much larger colony of fruit-eating bats. Had I not been present, the fruit bats would have been killed while the vampires escaped notice.

Finally, I was able to explain how to distinguish vampires from all other Latin American species. Vampires never have visible nose-leaves (a triangular-shaped cartilage that projects above a bat's nostrils) or tails. In contrast, all fruit- and nectar-eating bats of Latin America have easy-to-see nose-leaves and all insect-eating bats have equally visible tails. Furthermore, vampire roosts can be identified at a glance. Their feces look like reddish black tar, the result of an all-blood diet. Once the eradication specialists understood how to identify common vampires and that in fact most bats were beneficial, they were grateful for my help.

Unfortunately, similarly indiscriminate campaigns were being waged throughout Latin America, including with dynamite and long-lasting poisons in caves, killing millions of valuable bats and permanently destroying their roosts. In the 1960s, most Latin Americans used the word *vampiro* instead of *murciélago* for all bats. In fact, as recently as 1995, when Eugenio Clariond, an influential businessman and philanthropist from Monterrey, Mexico, joined Bat Conservation International's board of directors, his secretary dutifully updated his résumé to show that he was a trustee of Vampire Conservation International!

Lurid media exaggerations have grossly misrepresented vampire bats. They have never descended on humans in attacking hordes, as described by tabloids. Quite the opposite, they are timid and so small that I can hide one in the palm of my hand.

Since vampire bat diets do not permit fat storage required for hibernation or long-distance migration, they are restricted to living year-round in warm tropical or subtropical climates. Physiologist Brian McNab studied vampire energetics and concluded that if these bats attempted to live in a temperate climate, they would require twice as much food per meal, explaining why they are not found where winter temperatures fall more than briefly below 50 degrees. This apparently explains their current range from the lowlands of northern Argentina and Chile to northern Mexico.

Several thousand years ago, during a period of much warmer, tropical climate in North America, vampires that are now extinct lived from California to Florida and as far north as West Virginia. They apparently failed to survive when climates became cooler. Theoretically, because of climate change, vampires may be capable of a gradual return to the southern United States. They currently live just 130 miles south of our border. Their arrival would be a nightmare for conservationists, given the heyday tabloid journalism would have needlessly scaring people. However, there is otherwise little reason for concern. Vampires that feed on livestock are easily controlled, as I'll describe later, and incidents of biting modern humans would be extremely rare.

Of the three vampire species, the common vampire is by far the best studied. This bat has excellent night vision for finding prey during dark, moonless times when it is safest to hunt. Its highly specialized ears enable it to hear the breathing sounds of sleeping animals, and heat-sensing pits on its nose help find an ideal location where rich capillaries can be reached painlessly.

It also has razor-sharp incisor teeth that lack enamel in order to permit self-sharpening.

Since blood vessels constrict and blood clots rapidly to prevent excessive bleeding, a vampire must make a shallow incision of one-eighth- to one-quarter-inch in diameter and use its grooved tongue to continually lick the surface, injecting saliva with at least three ingredients that promote bleeding. An anticoagulant inhibits clotting. A second ingredient prevents blood cells from sticking together, and a third reduces constriction of blood vessels near the wound. As this potent saliva cocktail flows down the tongue's upper surface, blood is drawn up the underside of the tongue along unique grooves. Vampire saliva has even evolved to address differences in the blood of mammals versus birds. Whereas the saliva of common vampires is adapted primarily for mammals, that of the obligate bird-feeding white-winged and hairy-legged vampires is specialized for birds.

Vampires also have facial hairs that are in constant contact with prey, immediately alerting them to any movement that could indicate a victim's wakefulness and potential danger. Common vampires have, for their size, the longest thumbs of any mammal, and these serve as feet, enabling them to walk with stealth when stalking prey on the ground or to run with agility if attacked. Long thumbs also provide extra leverage to instantly catapult this bat into an airborne escape.

Bill Schutt found that bird-feeding vampires stalk perched prey, taking one slow upside-down step at a time along the underside of the branch where a bird is sleeping. Thus, not surprisingly, they don't need the common vampire's famous ability to run and jump, and they lack its extra-long thumbs. Schutt also observed that these bats sometimes snuggle up beneath hens in a manner that apparently causes them to be mistaken for chicks.

He credited the common vampire bat with an "unmistakable air of intelligence."

For those of us who have experienced vampire bats as the wonderful creatures they really are, it is difficult to accept the idea that the common vampire has truly overpopulated and requires lethal control in many parts of Latin America. Nevertheless, it is inescapable that those of us who care about bats must help ranchers, frontier *campesinos,* and indigenous peoples to minimize problems if we are to save millions of highly beneficial bats from eradication.

While helping Dieter Plage film vampires in Costa Rica, I met Dr. Hugo Sancho, a veterinarian who served as deputy director for animal health and protection in the Ministry of Agriculture and Livestock. He was deeply concerned about the impact of misguided vampire control that was inadvertently killing countless thousands of beneficial bats and wasting limited resources. He offered to give me a firsthand look at the problem if I could spare a few days to travel with him in the field.

Hugo was a handsome man with dark wavy hair and a ready smile. We drove for three hours through lushly forested countryside and small towns from San Jose to reach the Pacific coast town of Jacó. It was the beginning of the rainy season, hot and humid on the coast. We met a local veterinarian who had requested assistance with vampire control. From past experience, Hugo was dubious. He had already explained to me in Spanish, "Most of our requests for vampire control don't involve real vampires, but we cannot be sure without looking. It wastes a lot of our time, and if we don't quickly investigate, local people waste their limited resources killing beneficial bats."

Our first several stops were road culverts in the surrounding ranching country. We found lots of small fruit-eating bats but no vampires. As in my first experience many years earlier in Venezuela, we periodically spotted cattle with bloody necks, a sure sign of recent vampire bites. Hugo told the veterinarian, "Yes, you do have a vampire problem, but simply going around killing bats in roosts is a waste of time." *"Por qué?"* One of the ranchers wanted to know why, so Hugo explained, "Because the largest bat colonies are bats that help control insect pests, such as mosquitoes, or fruit- and nectar-eating species that are needed to carry seeds and pollinate flowers essential to forest health. Most vampires live in small, difficult-to-find colonies where we seldom find them."

The next question was, "So how can I protect my cattle?" Hugo grinned and said, "There is a much easier way. Can you herd your cattle into a corral before sundown? If you can, we'll show you tonight." The rancher was clearly curious. *"Sí, puedo,"* he responded in the affirmative.

After a quick trip back to town to reserve lodging for the night, we returned to the ranch, arriving a half-hour before sundown. Sure enough, the rancher had rounded up his small herd and was eagerly awaiting our arrival.

After a brief exchange of greetings, Hugo opened the special bat-catching box in the back of his pickup and began extracting equipment – four sets of extendable aluminum net poles, four 42-foot-long mist nets, nylon cords and stakes, a couple of electric headlamps, a pair of heavy leather gloves, and a homemade cage with rubber flaps on top for temporarily holding captured vampires. Last, he carefully extracted a jar of what appeared to be simple Vaseline. But he warned, *"Es muy venenoso"* – it's very poisonous. It contained a mixture of powdered rat poison, known as warfarin, an anticoagulant that can be absorbed directly through one's skin if handled carelessly.

The local veterinarian and the rancher couldn't resist touching one of the finely threaded nets, obviously wondering how such fine material could be used to catch bats. They were even more astonished to see us begin to unravel a ball of material no larger than two human fists to cover an area 42 feet long by 7 feet high along one side of the corral. Soon we had all four nets set, forming a box shape around about two dozen cattle.

Hugo had planned this trip to coincide with the latter half of the full moon phase of the lunar cycle. At that time the bright moon doesn't rise for several hours after sundown. He explained, "Vampires don't like to hunt when there is moonlight, so to most effectively catch them, it is best to set nets on nights when moonrise will occur at least two hours after sundown, but by midnight." That way the bats would be forced to arrive at a convenient time for us to catch them without staying up too late.

About an hour after sundown, as full darkness set in, we became quiet, barely moving or whispering so as not to frighten approaching vampires. Hugo had brought two small folding stools, and we each sat on one at the junction of two nets, so we would know immediately when a vampire got caught. We certainly didn't want to deal with 20 captives all at once, as I'd naively had to do during my first experience in Mexico.

Soon the first vampire struck one of Hugo's nets. Grinning from ear to ear, he encouraged the rancher and his veterinarian colleague to watch as he carefully removed it. Before putting it in his cage, he explained that vampire bats lacked visible tails and conspicuous nose-leaves.

Next, I caught a black myotis (*Myotis nigricans*), a tiny insect-eating species, and we used it to illustrate the tails of such bats and to explain that it fed on mosquito-sized insects. Our hosts were impressed to learn that it could potentially catch 1,000 insects in just an hour. Over the next hour and a half we captured five additional common vampires, each one placed in Sancho's cage. As the moon began to rise, we knew that there was little hope of catching more vampires, so we closed the nets and folded them back into their plastic bags.

Knowing what special animals they were, I hated having to watch Sancho paste each one with his vampiricide mixture prior to its release. The technique is extremely effective. When pasted individuals returned home, their roost mates would try to help groom the mess off, killing the entire colony, up to 40 for each one treated. As inhumane as it sounds, where such help has not been provided, ranchers have permanently destroyed thousands of roosts and millions of highly beneficial bats.

The next day, after an early breakfast, we set out for a larger ranch, again accompanied by our veterinarian colleague. This time there was a big difference. The veterinarian colleague had completely changed his thinking about bats.

Hugo and I stood back and smiled at each other while our colleague explained that most bats are beneficial and that by inadvertently killing the wrong bats in their roosts we would waste time and resources and create additional problems. Again that night, we demonstrated an effective solution.

Liked and respected by everyone we met, Hugo was a fount of knowledge about vampires and a wonderful spokesman for bats. On the long drive back to San Jose, he pointed out the obvious. "There's a big need for education. I already have Bat Conservation International's slide show on rainforest bats, and it is very popular. But what I really need is a program that specifically deals with vampire control and why it's important not to just kill all bats. People need to see for themselves why bats are important. I'd be happy to help you produce and promote an appropriate program."

Thus an idea was born – wouldn't it be great to produce such a program designed and narrated by Latin America's leading vampire control experts? On returning home, I quickly called my friend and colleague Rex Lord, who was serving as an expert on vampires and rabies control in Venezuela at the time. I asked, "If Bat Conservation International can find sponsors, would you be willing to collaborate with Hugo Sancho of Costa Rica in producing a vampire and rabies control training film for veterinarians and ranchers?" His immediate answer, "Yes, it's badly needed. When can we start?"

Six months later, Rex and I teamed up with Hugo and a film crew in Costa Rica. They made a powerful partnership, explaining the latest techniques and the importance of avoiding harm to other species. The 26-minute video, titled *Control del Murciélago Vampiro y La Rabia Bovina*, was completed in 2002 and has been widely distributed.

Thanks in large part to the dedication of Rex and Hugo, combined with wide use of our program by Latin American bat research colleagues, massive programs of indiscriminate killing are now a thing of the past. Government personnel are increasingly becoming leaders in bat conservation, working diligently to educate growing numbers of ranchers and *campesinos* regarding the advantages of targeting only the vampires that feed on livestock.

Nevertheless, when I collaborated with the U.S. Forest Service International Programs in 2009 to provide a training workshop for Paraguayan leadership personnel, we quickly demonstrated the need for much more education of poor people in remote areas. We led a dozen participants through the frontier village of Bahía Negra on the border with Brazil, looking for bat roosts in hollow trees.

Virtually every tree capable of sheltering bats showed evidence of fire or other signs of bat killing. Villagers explained, "*Matamos vampiros.*" (We're killing vampires.) However, when we examined a dozen burned-out hollows we found evidence only of insectivorous bats. The lone vampire we discovered had gone undetected in an old military barracks. Residents readily agreed to protect their remaining bats, once they realized they were helpful.

Despite the nearly universal notoriety of vampire bats, they seldom pose a threat to humans, constitute only a tiny fraction of our planet's bats, and can easily be controlled without harm to other species. Grossly exaggerated myths aside, vampire bats rank among our planet's most fascinating, even valuable animals. And they may even one day save your life!

Vampire saliva contains a veritable treasure trove of unique regulatory molecules of potentially great value in the development of modern medicines. Just one of the substances discovered from the common vampire, desmoteplase, is currently being tested as a much-improved treatment for human stroke victims.

CHAPTER 5

BATS THROUGH A CAMERA'S EYE

"OH NO! please don't even think of using those pictures." I had come to Washington, D.C., to meet with book editor Tom Allen at the National Geographic Society, and he had just shown me the proposed layout for my chapter on bats in a book titled *Wild Animals of North America*. It was to be published in their Natural Science series.

A bit surprised, Tom asked, "What's the matter?"

I explained, "I've worked hard to counter myths and unfounded fear in my chapter, but these pictures will undo everything I've said." Most of the photos showed bats snarling in self-defense, with bared teeth.

He admitted that they did look vicious, but commented, "We've tried hard to locate better images, but these are the best we've been able to find. Do you know of any better ones?" I didn't, so Tom offered to send Bates Littlehales, one of National Geographic's staff photographers, to the field with me to see if he could get some better shots.

It was June 1978 when Bates accompanied me to photograph gray myotis at my eastern Tennessee study site in Oaks Cave. He wasn't exactly enthused as we slithered on our bellies through a narrow entry passage carrying a tripod, several flash stands, and his equipment bag.

The gray myotis colony had just emerged to feed, so we'd have about an hour to photograph newborn young before mothers began to return. It took 15 minutes just to reach the roosting area. The cave was wet and muddy, and Bates's flash stands were designed for flat studio floors, not uneven rock. The idea of needing flash stands and multiple flashes was new to me. I didn't like it, especially when one of his stands fell over, dumping an expensive flash into a soggy mixture of mud and bat guano. Expletives filled the air as Bates fished his sinking flash from the gooey mess and prepared a spare.

In the nick of time, he got set up with me holding his second flash in place. Through a telephoto lens, he clicked off several shots at a variety of exposures. He would have liked to try additional angles, but I insisted on repacking and leaving in a hurry. We didn't want to be seen by returning mothers, who might move their pups to a less desirable roost if disturbed. One year, when naive spelunkers had disturbed them, the colony had moved its young to a cooler roost where they couldn't grow fast enough, and many had died.

Next, we rigged a special photography set over the Norris Reservoir in hopes of showing a gray myotis catching a mayfly over a hatch site that several females guarded as a feeding territory. We rigged flashes on poles protruding above counterweighted buoys 20 feet from shore, and we had an additional bank of flashes and a camera with telephoto lens on the shore. We spent hours setting it up, carefully calculating flash exposures. Then we waited expectantly for the bats to arrive. Unfortunately, just as the bats came close and all seemed perfect, a powerful motorboat passed and ruined our set.

For nearly a month, we tried a variety of ideas. Bates's large, high-speed flash set was simply too bulky to be used in the cave where we wanted to take flight shots during an evening emergence. It was extremely difficult to find just the right location, even for small flashes. We needed a site narrow enough to funnel flying bats where they would predictably break his infrared beam, thereby triggering three flashes.

In order to avoid using the bulky flashes, Bates came up with a new idea. Since we were working in close quarters, perhaps he could dial down the power to a fraction of normal on his regular flashes to make them high-speed. He then positioned them within two and a half to three feet of where we hoped a flying bat would strike the beam. The camera was mounted on a tripod and carefully framed and focused on that location. He would hold the camera shutter open with a cable release until a bat struck the beam, then allow the shutter to close after the triggered beam fired the flashes.

Each photo covered only a small area, meaning hundreds of shots would be required in order to get a bat properly positioned in the frame. It was great fun, sitting on cool rocks, watching the flashes fire. Each time, we could see the exact location and wing position of the bat and get a pretty good idea if we'd like the resulting picture. Although three flashes were firing in only about a 10,000th of a second, Bates's light-sensitive "slave" units could synchronize the firings sufficiently to avoid multiple images on a single frame.

Fortunately for the world of bats, Bates was kind enough to explain his multiple flash lighting and speed-stopping techniques to me. He could hardly go to the restroom without me asking how and why. We became friends and enjoyed working together. Nevertheless, in a whole month, he got only half a dozen great shots, far fewer than we needed to go with my book chapter. Clearly, bat photography was going to be challenging.

On his departure, Bates said, "Merlin, I can see that success depends as much on understanding bats as on knowing photography. Now that you've learned all my secrets, why don't you purchase a little more equipment and see if you can't get some shots on your own. Here, take my extra film." And so began the world's largest collection of bat photographs, a key step toward the founding of Bat Conservation International, and a revolution in public perceptions of bats.

A short time later, illustrations editor Anne Kobor called and asked if I could tell Bates where to find a Mexican fishing myotis (*Myotis vivesi*) to photograph for the book. I explained that it would be difficult for him to do that on his own and volunteered to accompany him as his field assistant. I was enthusiastic about the possibility of learning more from him. Instead, Anne called me back and said, "I've spoken with Bates, and he says that if you are willing to go, he won't be needed." I was shocked, but she reassured me, explaining, "We don't need any high-speed action. Just a nice shot of a perched bat eating a fish will be fine."

I called Bates, and his advice was, "Don't worry about the expenses as long as you get the picture. Just don't come back without it."

As one who already had been frequently accused of overkill, I took this advice seriously. Prior to my departure, I shot ten rolls of test film, experimenting with widely available flashes to stop high-speed action with reduced power and to determine the best light ratios. I also called every colleague I could find who might have had experience with Mexican fishing myotis. From the scientific literature, I'd deduced that these bats would be relatively easy to capture by hand, since they often lived under rocks on islands in the Sea of Cortez. But my colleagues' advice was discouraging: no one had ever gotten one of these bats to eat in captivity, my first priority for success.

My then wife, Diane Stevenson, and I rented a beach cottage and a boat with an outboard motor in San Carlos Bay and went in search of the bats. To our dismay, we could find no reachable bats. Though present on nearby islands, the only bats we found were roosting high up in cliff-face crevices, protected by breakers that threatened to crush our small boat on sharp, barnacle-covered rocks.

Nevertheless, remembering Bates's advice about not coming back empty-handed, we approached the nearest island. Using the outboard motor, Diane held the boat a hundred feet out from danger, while I swam to protruding rocks below the cliff. I did my best to ride the tail end of a wave instead of getting caught and dashed on the rocks. But the waves that hadn't looked bad from a distance turned out to be six feet tall, with only short intervals between. By the time I reached the cliff, I was cut and bleeding profusely.

I began what seemed like a fairly simple climb from ledge to ledge to reach a roost some 40 feet up. Nevertheless, wearing only a wet bathing suit, I was forced to abandon my mission ten feet short of the bats. The cliff face was coated in guano from roosting sea birds, and the water dripping from my body made the going treacherously slick. The climb down was a terrifying experience, but an even greater terror waited at the bottom.

As I crouched just above the breaking waves, I first noticed that submerged rocks would prevent me from diving in to get under and beyond the next wave, and without that head start it would be difficult to avoid being thrown back onto the rocks. I waited as long as possible before marshaling the nerve to jump feet first into a narrow space between two boulders and begin the fastest swim of my life.

Just as I got safely past the waves and could look up, I discovered that Diane and our boat were no longer where we'd planned to rendezvous. Another quick look and I saw her valiantly attempting to start the stalled motor as she drifted perilously close to the breaking waves.

As I treaded water in the open ocean, I recalled having seen a shark fishing village, and visions from the movie *Jaws* filled my mind. Even if Diane jumped overboard in time to save herself, I held out little hope of our swimming through more than a mile of ocean to reach shore, given that I was still a bloody mess that should attract any sharks in the vicinity.

Just as I was imagining what the first shark bite would feel like, I heard the sputter of an engine. Within seconds of disaster, Diane had somehow gotten the motor started and was on her way to get me. I've never scrambled into a boat faster!

That night, we traveled miles of coastline, using a powerful spotlight to search for feeding fishing bats. We found them in only one place where we might be able to set up a mist net to catch them. The next evening, we set a 42-foot net low over the water and waited patiently on shore with our night-vision scope. After roughly an hour without seeing a single bat, suddenly a two-foot Mexican needlefish jumped into our net and began rapidly dragging it into the bay. I cursed our bad luck and dashed into the water to rescue the net. To our great surprise, the struggling fish apparently attracted the fishing bats we were looking for. We caught seven all at once.

We now faced another big challenge. These bats were incredibly gentle, but as we had been warned, they refused to eat anything we offered, including freshly caught minnows. To overcome their fear, we spent many hours carrying them around, stroking and talking to them. Yet, 24 hours later, none had eaten and we feared failure. Unless we could convince the bats to eat, we'd soon have to release them without photos.

Our big breakthrough came when I tried cutting a minnow into tiny pieces, offering just one piece at a time. I slipped a piece between the lips of one of the bats, and moments later the bit of fish had disappeared. Adding another, I watched carefully, breathing a huge sigh of relief when I saw a tiny tongue tip take the offering. At first, only a couple of bats would accept the tiny pieces. We kept giving them larger and larger offerings until finally they accepted whole minnows. Then we tempted the others by arranging them within inches of the two that were now greedily eating. Soon, they all turned ravenous. We had learned the first and biggest secret to successful bat photography: overcoming fear is paramount, and the best way to a bat's heart is through its stomach.

We spent the next morning gathering cliff-side rocks to build my first photographic set. Finding rocks that looked like a cliff face was challenging. We had to collect fallen cliff-face pieces that weren't too heavy to be hung together in a natural-looking set.

As we were returning with our trophies at lunchtime, we heard a woman screaming, then realized the sound was coming from our cottage. It was our maid. The door crashed open as she ran headlong into me, screaming, "*Vampiros, vampiros!*" We'd forgotten all about maid service and what might happen when seven bats with two-foot wingspans were encountered loose in the bathroom. We had arrived just in time to explain, though that was the end of maid service.

By that evening our now tame bats performed perfectly on our set. We took more than a hundred apparently great photos showing them hanging by their huge feet, eating minnows. These bats were related to the gray myotis that had first caught my attention. They belong to the same genus, and both have large feet for snatching prey from the surface of water. But the feet of Mexican fishing myotis are truly enormous and have much longer claws. Also, both toes and claws are flattened to slip through water with minimal resistance. Adaptations for fishing have evolved several times independently in the genus *Myotis,* including in species I have subsequently photographed in Australia, Southeast Asia, and China.

On our final night in Mexico, we tried for a surprise bonus. We set up the flashes on stands with an infrared beam triggering device – just as I had practiced at home – and coaxed one of our bats to fly through it carrying a minnow in its mouth. We appeared to have gotten a couple of potentially good shots. In those days, of course, we didn't have the luxury of the instant gratification now available through digital photography. We could only hope we had what we wanted.

Later, we gave our bats a banquet feed and bade them farewell, hating to see them go. In the morning we returned home and express-mailed our unprocessed film to National Geographic.

Weeks later, when Anne Kobor called to congratulate me on what she reported to be "one of our all-time-best nature photos," Diane and I were extremely pleased and relieved. In fact, when we saw the picture, we were amazed. One of the shots showing a bat in flight with a fish had turned out perfectly. Encouraged by such initial success, I continued to photograph bats and became the second most used photographer in the book. Soon I began to see just how much impact good photography could have in educating a skeptical public to appreciate bats.

Thrilled with the progress, I still had much to learn before becoming a consistently successful bat photographer. While trying to recapture fishing bats flying around loose in our San Carlos Bay cottage, it had become apparent that a collapsible, walk-in studio would be enormously helpful.

I contracted with a tent and awning company in Milwaukee to build me one out of tough mosquito netting sewn together with canvas seams and a zipper in one corner. Canvas flaps at the bottom prevented escape, and Velcro in the corner farthest from the entry allowed for rapid attachment of black velveteen background material. The studio was ten feet square by seven feet tall, held upright on four extendable aluminum poles. An enclosed area can be extremely helpful in working with bats.

"What? Are you joking? You really want me to move all the furniture out of a guest room so you can keep bats in it?" The manager of the Golf Kakamega Hotel in Kenya was responding to my request to reserve three rooms, one for myself, one for my two assistants, and one for my epauletted fruit bats. He'd been happy until I mentioned clearing a room for bats. Having anticipated such a response, I'd come prepared. I calmly showed him my frog-eating bat article in *National Geographic* and explained that I was working on assignment.

Suddenly, his entire demeanor changed. He smiled from ear to ear and said, "How soon do you need the bat room to be ready? Just let me know how I can help. We're at your service."

It was 1984, and we were at a resort hotel in the mountains above Lake Victoria, Kenya. I was planning to spend the next month photographing epauletted bats for an article in *National Geographic* magazine.

An hour later, the furniture had been removed, and I was covering the carpet with plastic sheeting. Next we unpacked my studio and a staple gun and began stapling studio guy lines to wood paneling. These would hold the four extendable corner poles in place when we raised the enclosure. It was ready for occupancy in two hours.

At dusk, my helpers drove to nearby gardens and set two 42-foot-long mist nets around flowering banana plants. They returned with four little epauletted fruit bats (*Epomophorus labiatus*), three females and a male. These were the first of a cast of more than a dozen small flying fox–type bats that would become stars in my April 1986 *National Geographic* article titled "Gentle Fliers of the African Night."

Weighing less than two ounces each, these bats were so small they could almost be hidden in the palm of my hand. Like other epauletted species, the males had inflatable cheek pouches as well as shoulder pouches in which were hidden epaulettes of long, white fur that they could flash in a showy "song and dance" display to attract females. Both sexes had large eyes and cute faces, and uniquely sported a white spot above each eye.

A half-dozen or more epauletted species live in tropical and subtropical Africa, where they are important seed dispersers and pollinators. I planned to use these cute little bats as ambassadors for flying foxes in general, all of which belong to a single family, the Pteropodidae.

I would rely on the portable studio to get close-up shots of mothers and their pups interacting in settings arranged to be indistinguishable from those encountered in the wild. Here my bats could also be photographed feeding on rainforest fruits and flowers that otherwise would be unreachable high up in the forest canopy.

Described in such simple terms, it may sound easy. It wasn't! It just made the impossible barely possible. I had a real scare the very first night. When we released the first four bats into the enclosure, despite my best efforts and much patience, they refused to calm down or accept food from my hand. I tried repeatedly from ten in the evening until four in the morning. Fruit- and nectar-eating bats normally run out of energy, get hungry, and are among the first to cooperate, but, unlike the frog-eating bats I would learn to train in Panama, these just would not give in.

It was sobering to think that *National Geographic*'s senior editor for research grants, Mary Smith, had already wired me a $20,000 cash advance based on a single long-distance phone conversation from Nairobi. I had assured her that, if she could cover my expenses, I could produce an article about epauletted bats that editor Bill Garrett would love. Now, I wondered if my assurances had been premature.

Determined not to give up, I decided to take an hour's nap, allowing the bats to get hungrier. When I returned, all four were lying on the floor, apparently dead. Their bodies were cold to the touch, but when I picked them up I realized they were still alive.

I immediately rigged up a lamp to warm them and filled a syringe with sweetened mango juice. I tried putting a drop in each of their mouths. Gradually, their tongues began to move. Their bodies warmed, and they began to eagerly lap the juice. Within about 30 minutes, all four were back to normal, and I knew they'd be fine. The only difference was that they no longer feared me and would readily eat chunks of banana from my hand.

The next evening, my assistants set nets at a fruiting fig tree in the Kakamega Forest and caught a male and two female Wahlberg's epauletted fruit bats (*Epomophorus wahlbergi*), a species similar to the little epauletted fruit bats except for being twice as large. Unlike their predecessors, these readily accepted food from my hand. Night after night, I virtually lived with my bats until I could call any one of them to my hand, pick them up, carry them around, even wipe a dirty face with a tissue. As with all bats, each had its own personality and intelligence, and some would permit liberties not accepted by others. Knowing their individual personalities was essential.

I hired a carpenter to build wooden stands strong enough to support heavy branches for set building, paid a local boy to clean the plastic-covered floor, and contracted with a vendor to deliver approximately five pounds of ripe bananas and papayas every afternoon.

Within a week we could begin work on a list of essential shots that I had planned in advance. We would work for days at a time on single photos, each requiring a different set. To prevent leaves from wilting in a hot climate, we had to submerge freshly cut branch stalks in sturdy plastic bags filled with a mixture of water, sugar, and detergent. I would often spend hours searching for branches with just the right contours for a pleasing image, natural down to the finest detail.

One of our greatest studio challenges was to obtain a photo of an epauletted bat taking a ripe fig. We could see dozens at a time taking figs in a large wild fig tree in the forest. Most of the bats, however, were at least 50 feet up in the canopy. Also, it was rare to spot a full cluster of colorfully ripe fruit. Most had been picked over, and those that remained weren't photogenic.

When we finally spotted a beautiful cluster of ripe fruit, it never occurred to me to suppose there might be a very good reason why it hadn't been touched. One of my assistants was an expert climber and didn't hesitate to climb up for it. To be sure that this rare find wasn't damaged, he carried clippers, two long cords, and a small saw. The cluster would have to be detached from its branch, carefully tied and lowered independently, then reattached back in the studio.

Almost immediately when he arrived at the figs, my helper began screaming in agony. He was being attacked by a horde of large ants. But he had a lot of pride and was determined. In between swatting ants, he managed to tie off the fig cluster and quickly lower it to me. He even stayed long enough to cut and lower the branch. The instant he hit the ground, he did the fastest strip-down ever seen. He was covered in bites, and his clothes were saturated with ants. We spent the next half-hour getting rid of them.

It took hours to organize the set and prepare the bats. First, we brought additional branches into the studio and mounted them on stands such that, looking through my camera, it would not be possible to tell that bats taking figs weren't out in the rainforest canopy. Additionally, none of the branches could block backlight flashes needed to clearly outline approaching bats. And prior to bringing our precious figs into the set, we had to hand-feed all but one of our bats to the point of satiation. That way we wouldn't have to contend with a dozen bats trying to take figs all at once. I chose the smallest bat, a juvenile, to be the one allowed to take figs from the cluster. Because of its small size and light weight, it was least likely to inadvertently damage figs other than the ones it took. Miraculously, when the set was complete, none of more than a dozen ripe figs had been jarred loose.

After taking meticulous light readings, I preframed and focused the camera, and made certain that an assistant was constantly guarding the figs against any of the bats prematurely attempting to take one.

At last we were ready. An assistant carried the young bat

to a location from which its approach would be appropriate to my framing and hung it on the studio netting. I was concerned that our bat would approach from the wrong angle and knock off several figs before I could get even one good shot. We waited in suspense for several minutes. A large male attempted to come and was chased away. One of my helpers suggested, "Maybe the young one isn't hungry."

And I responded, "Let's wait a little longer."

Ten minutes later, our designated pup began to circle the studio. That wasn't part of the plan, and I nearly intervened. At that moment, the bat did a perfect approach, grabbed a fig, and carried it away without damaging any others. I got two apparently ideal shots as it left with a fig in its mouth, breathing a huge sigh of relief that all had gone well. Soon our pup returned for seconds, and eventually others joined the feast. Assuming my calculations were correct, we should have photos to spare.

Spending night after night with my bats, I noticed that mothers seemed to be teaching young pups the scents of novel fruits by allowing them to lick their lips after they had fed. Knowing that would be an extraordinarily winsome shot, I organized a special set, placed the mama I'd most often seen exhibiting this behavior in it with her pup, then handed her tidbits of novel fruit in an attempt to induce a repeat in front of my camera.

She and her pup obliged, but the action was so quick that, if I waited until I saw the pup licking, I'd be too late. Most film cameras had a delay of about 1/20 of a second from the time of triggering to the shutter actually opening. Time after time, I shot and missed. I had to anticipate the action. It took a total of 17 hours, but I finally got the shot.

Some things just can't be shot in a studio. To complete the story, I needed two key photos. To show a whole colony in a natural roost, we had to drive for two hours down to the Kaloka Veterinary Research Station on Lake Victoria. There, a colony had learned to ignore humans in an area where they were safe from harm. Nevertheless, getting the shot was far from easy.

I needed to balance flash illumination of the bats with a blue sky overhead. Morning after morning for nearly a week, we had to leave at 2:30 in order to arrive and get set up in time to shoot against an early morning sky. And five times in a row, the sky was briefly obscured behind clouds during the 20-minute span of appropriate lighting.

Photographing a courting male with a female hovering in front was one of the most difficult challenges of my entire photographic career. In the wild, these bats are very shy. At the veterinary research station, I had noticed that male epauletted bats courted only when light levels were sufficient for females to see their flashing epaulette display, meaning at dusk and dawn, except when there was a moon.

While driving through the town of Kisumu with my windows down, I had noticed an apparent exception. Males courted late into the night where they could be seen in the light of streetlamps. In fact, they guarded well-positioned lamps as courtship territories. Armed with that knowledge, we searched for a location where the bats might have grown to ignore humans below.

A large shopping center proved ideal. We found a male calling near an entry point, and I drew a detailed map of exactly where he consistently hung to sing and perform his wing-dance. The next afternoon, we arrived early enough to jockey for an ideal parking space for our Land Rover, and I set up my camera with a 300 mm lens and a sturdy tripod on top of the vehicle. I consulted my hand-drawn map of the bat's anticipated courting location, then raised a film carton on an extendable aluminum pole to where the bat would be and focused the camera on it. We planned to raise two flashes on 20-foot poles, these to be hand-held by my assistants. They had to be positioned and aimed exactly right.

The first night our bat got nervous and moved to an alternate perch. He didn't like our flashes weaving around in the vicinity of his previously used location. I sent an assistant to hold up a long pole near enough to the new perch to make him nervous, and he returned. After repeating this several times, he gradually realized that he would have no peace until he came back to our location. As he resumed calling, we waited expectantly for females to arrive.

None came for more than an hour. Then we sensed something was about to happen when our male rapidly increased his gonglike calls to a staccato. The change reminded me of a Geiger counter when radioactive material comes near. His wingbeats became a virtual blur and his voice louder. His enthusiastic expectancy was clear. I looked to the sides to check flash distances and aim. Neither was right! There was no time to explain, so I didn't bother to shoot when a prospective mate briefly hovered in front. Moments later, she departed, apparently having found our suitor inadequate.

I quickly explained to my assistants that I couldn't shoot unless both their flashes were held just right, and we waited for the next female to arrive. When one did, the flashes suddenly lurched forward, scaring both bats. Further explanations ensued, and we waited some 15 minutes for our male to return. Soon after, he alerted us to another approaching female. By this time, a light breeze had sprung up, causing our perched bat to sway up to a foot from side to side. All went well except that he had been blown out of my frame. That would become our most frequent problem. Other shots that could have been perfect were spoiled because one of the bat's wings happened to be between the other bat's face and my camera. By two in the morning, I realized that we were in for a really tough time.

We packed our equipment and began a roughly two-hour drive back to our hotel. The next afternoon we were back again. After a week of this, our biggest problem was staying awake. We gradually accumulated a store of "possibly good" shots, but some 660 frames later, we still couldn't be absolutely certain, and this was the anchoring picture for my story.

When I finally returned to Nairobi to catch a flight home, I had shot more than 10,000 photos without having seen a single one processed. In the modern age of digital photography, it's difficult to even imagine the stress of having to shoot thousands of images without getting to see a single one until weeks later. Back at home, once the processed film came back, I immediately went to the rolls containing courtship images and began to feverishly sort for the one great shot I needed. I was like a drug addict in desperate need of a fix.

After sorting through more than 600 images, I'd seen sufficient near misses to know it was possible, but I still didn't consider a single one adequate. Hands trembling, I scarcely dared look at the last half-dozen. But there it was – *the shot*.

Why go to so much trouble to get such difficult photos? Because that's what it takes to get into *National Geographic*, and there is no substitute for such credible worldwide publicity when one is promoting conservation of animals as unpopular as bats were at the outset of my career.

This renown proved vital when, on my return home from Africa, I learned that Judges Cave, one of the most important bat nursery sites in the southeastern United States, was about to be buried under a proposed housing development. When I called the owner and explained its significance, he obviously thought I had to be some crazy kook. He pointed out that his home was nearby, that he had children, and that he knew bats carried rabies. He promised, "I'll have it bulldozed shut tomorrow and save you any further concern."

Desperate, I asked, "Do you ever read *National Geographic?*"

He proudly responded, "Yes, we're members."

I questioned, "Do you happen to have their recently published book, *Wild Animals of North America*?" When he answered in the affirmative, I asked him to please read the chapter on bats to see what the National Geographic Society had to say. I promised to call him the next day, counting on him to notice I had authored their chapter.

The next morning when I called, I'd quickly gone from crazy kook to esteemed National Geographic scientist. "I'm sorry I didn't recognize who you were. Of course, we won't do anything without your advice. My family is still a bit nervous about bats, but they'd probably listen to you. Could you come down and show us our cave and explain about the bats?"

Armed with a plane ticket paid for by Verne and Marion Read, friends of the Milwaukee Public Museum who funded many of my early conservation efforts, I was in Marianna, Florida, a few days later.

My Florida host was cordial and enthusiastic. He introduced me to his wife and daughters, and we shared a delightful evening watching approximately 100,000 bats emerge from their cave. When I caught one and showed it to the little girls close-up, their immediate response was, "Oh, it's so cute!" The former bat hater ended up generously collaborating with the Florida Nature Conservancy and the Florida Fish and Wildlife Conservation Commission to permanently protect the cave and surrounding habitat as a bat sanctuary.

Time and again, photography has proven key to bat conservation progress, which is why I've spent so much of my career on it. My five articles in *National Geographic* and the *Secret World of Bats* documentary would eventually reach countless millions of people in more than a hundred countries, dramatically improving the public's perception of bats.

CHAPTER 6

DISCOVERING FROG-EATING BATS

EXCITED AS A CHILD AT HIS FIRST CIRCUS, I sat riveted to my night-vision scope. Three frog-eating bats (*Trachops cirrhosus*) were hunting at a small jungle pond on Barro Colorado Island, Panama. Hundreds of frogs were calling in an almost deafening chorus, and the bats were having a virtual picnic catching frog after frog, sometimes practically at my feet. These observations would provide the first documentation of a bat that specialized in hunting frogs.

Staring into a lens similar to a small television screen, I could clearly watch the action. I'd illuminated the pond with an infrared light, visible only through my night-vision scope, which magnified the light 64,000 times. The revealed scene was captivating. Rain earlier in the day had brought túngara frogs out in force. It was February 27, 1980.

Three years earlier, I had inadvertently caught a frog-eating bat carrying a partly eaten frog while helping my early mentor, Charles Handley, with his fruit-eating bat research. I was intrigued, wondering how a bat could risk eating frogs in a jungle where so many frogs of that size were poisonous. If a bat merely listened for sounds of hopping frogs, or detected them using echolocation (emission of sound pulses, using returning echoes to gain information about prey or the surrounding environment), how would it know which ones were edible prior to grabbing them and being potentially poisoned?

The logical answer was to hunt only males that were courting, since, like birds, each species had its own unique calls. Indeed, if bats could identify edible frogs by their calls, the pickings should be lucrative.

If I was correct in hypothesizing that bats could identify and catch frogs by homing in on their calls, it was possible that these predators were having as much impact on male frog courtship behavior as the female frogs were. Charles Darwin was the first to speculate on the possibility of such counter-selection from predators. But it still hadn't been demonstrated for vertebrate animals.

Intrigued by the possibility that a predacious bat could influence frog courtship behavior, I immediately submitted a grant proposal to test that hypothesis. It was rejected by reviewers who responded that it simply wasn't possible for bats to hear low-frequency frog calls.

At the time, I was employed as curator of mammals at the Milwaukee Public Museum in Wisconsin and convinced museum donors Verne and Marion Read to fund a two-week trip to the Smithsonian Tropical Research Institute's field station on Barro Colorado Island. They were an adventuresome couple who welcomed the opportunity to serve as my field assistants. Following their participation in my frog-eating bat research, Verne would go on to serve as the founding trustee of Bat Conservation International.

I hadn't yet been able to obtain a night-vision scope to make direct observations, so we had to search for circumstantial evidence. We had set mist nets where frogs were calling versus not calling. The nets where frogs were calling caught 36 times more fringe-lipped bats, a species destined to become known as the frog-eating bat.

Resident Smithsonian herpetologist Stan Rand had recordings of many species of frogs calling. I asked him to bring his tape to a flight cage where I had released one of these bats. As the tape played, the bat periodically hovered over the speaker while completely ignoring it at other times. I had no idea what it meant, because I didn't know any of the calls. When the tape ended, Stan exclaimed, "Wow! I think you're on to something. That bat ignored calls of frogs that are either poisonous or too large to eat, but seemed attracted to calls of potentially edible species."

Armed with this evidence, I presented my hypothesis at the 1979 annual meeting of the Society for the Study of Amphibians and Reptiles held in Knoxville, Tennessee. I explained my preliminary results, suggesting these bats were influencing the evolution of frog courtship behavior. The herpetologists were skeptical, noting that none had ever seen predators attracted to frog calls, despite the countless hours they'd spent studying courting frogs.

I asked how many of them had ever hidden quietly with a night-vision scope to make their observations. They laughed, realizing that none had. Frogs might pay little attention to scientists and their headlamps, but I reckoned that predators might.

Stan had studied the courtship behavior of túngara frogs 16 years earlier. He'd found that males were reluctant to use the calls that females found most attractive. Short, whine-type calls were far more attractive to potential mates if accompanied by one or more chuck-type calls, the more the better. Apparently, the extra chucks at the end of a male's call made him easier to locate.

Rand had hypothesized the existence of a predator that, like the female frogs, preferred the easier-to-locate calls that included chucks, hence the males' reluctance to use them. But no such predator had been discovered. By early 1980, with Rand's enthusiastic endorsement and a new night-vision scope, I'd gotten a research grant from the National Geographic Society's Committee for Research and Exploration and had returned to observe how bats responded to courting frogs. I was amazed to see so much action on my very first evening.

At dusk I had set up my scope, with a rheostat-controlled infrared light source, on a sturdy tripod. I'd selected a wide-angle lens so I could see the entire Weir Pond. Water covered an area about the size of two Ping-Pong tables, and it was conveniently located just 100 yards from the Smithsonian Research Institute field station's main facilities. By the time it got dark, I was seated quietly on a small folding stool less than ten feet from the water.

This was a known breeding site for túngara frogs. As I waited for the frogs to appear, I was delighted with all the nightlife such a pond could attract: first a philander opossum came to drink, then a South American bullfrog assumed a commanding position, followed by tiny insect-eating bats fluttering in for a sip. Somewhat disconcerting were the fireflies that appeared like fiery comets as my night-vision scope magnified their flashing lights by 64,000 times.

Almost immediately, I could hear the first cautious whine calls of male túngara frogs. These early callers appeared to be rather nervous. The first one gave a simple whine call, and waited several minutes before trying again. As numbers grew, the frogs became braver, answering each other more frequently. Their sexiest chuck calls were added only after a substantial number had joined to make a chorus.

Just then, three large bats arrived. Instantly, the entire chorus stopped. There was a bright, first-quarter moon barely visible through the forest canopy. Could it be that these mere inch-long frogs could recognize approaching frog-eating bats in time for an entire chorus of 50 or more frogs to shut down in a split second?

The bats simply spread out around the pond and found convenient perches in bushes near the water. Then they waited patiently. Were they waiting for the frogs to resume calling, as I suspected? One by one, the frogs began again with simple whine calls, gradually working back up to a full chorus, including some daring to use added chucks. The bats continued to wait for several minutes longer, but I could see their big ears rapidly twitching back and forth as if getting fixes on frog locations. One bat swooped low over the pond. There was a quick splash, and it returned to its perch with a frog in its mouth.

Again and again, the process was repeated. The bats actually appeared to take turns. By late evening, frog numbers grew into the hundreds despite the bats. They seemed to play what I thought of as "frog Russian roulette." With only a few bats hunting, the odds of any one frog being eaten became so small that they just threw caution to the wind and did their sexiest best.

By that time, I was completely enthralled.

I could hardly wait to tell Mike Ryan, a herpetology graduate student from Cornell University who was working under Rand's guidance. We'd met just a few days before, and since Mike was doing his doctoral research on the courtship behavior of these same frogs, he was enthusiastically hoping to help me test Rand's predator hypothesis.

A few nights later, we set a mist net among bushes near calling frogs, and sure enough, we soon caught a frog-eating bat. We handled the bat with care to avoid unnecessarily frightening it, put it in a soft cotton bag, and carried it back to our flight cage. Inside, I slowly opened the bag and stroked the bat. Then I offered it a freshly killed túngara frog, which it ate with gusto. As I gradually opened my hand, it flew, circling the cage before landing in one corner.

I approached slowly, holding out another frog, and the bat flew away. Next time it flew again, but allowed me to come a little closer. Breaking in a new bat takes patience. Nearly two hours later, as the bat got tired and hungry, it finally took the frog from my outstretched fingers. By the third successful approach, I began making soft clucking sounds as I came closer. The next night, when I held out my hand with a bit of minnow and clucked, the bat flew right to me for its reward. It had overcome its fear.

Now we were ready for our first controlled experiment. During the day, we had arranged two speakers, one each in opposite corners of our 15-foot-square flight cage so we could test our bat's preferences. In dim red light, we sat in the corner farthest from where the bat hung from the ceiling. Mike had made tape loops that would consistently play a túngara frog's lone whine call from one speaker, and a whine with added chucks, which female frogs preferred, from the other speaker.

The suspense was palpable as we began to play our calls. The bat twitched its huge ears, just like I'd seen at Weir Pond. Then it flew directly to and hovered over the speaker playing the whine-plus-chucks calls. We were both grinning from ear to ear. Mike turned off the tapes, and our bat returned to its perch in the far corner.

We switched the speaker wires and waited several minutes. This time the whine-plus-chucks call came from the opposite speaker. Again, the bat twitched its ears and went directly to the speaker playing the call with added chucks. After repeating this procedure a dozen times with the same outcome, we were all but dancing in our exuberance. On our first trial, we had obtained results that were so significant that reviewers of our first scientific manuscript would kid us about why we'd bothered to run statistics on the obvious. Rand's hypothesis from 16 years earlier had been confirmed; there was a predator that preferred túngara frog calls with added chucks.

But following that first success, disaster struck. Two nights later, when we attempted to test another bat, it just hung and wiggled its ears. It would occasionally fly around the enclosure, but it pretty much ignored our speakers as we played call choices. For nearly a month, we were unable to get even one more bat to choose between our speakers. We knew something had changed, but we couldn't figure out what it was.

In all of our experiments, we had laid the speakers on their backs, facing the flight cage ceiling. Each speaker was 11 inches long with separate woofers (for low frequencies) and tweeters (for high frequencies). At the beginning, woofers and tweeters had been aligned toward the corner where bats perched.

Finally, I realized that since that first experiment we had inadvertently changed speaker orientations. Without realizing it could make a difference, we had rotated the speakers so that the woofers and tweeters were no longer aligned toward the bats. They were now perpendicular to the perched bats. Even I considered this to be an unlikely cause for our problems, but desperate for a better idea, I pointed it out. "Maybe our call recordings don't sound natural when each call is clearly coming from two different locations in the speaker."

Mike laughed, agreeing that as desperate as we were, we might as well try rotating the speakers.

We did, and a previously unresponsive bat began to perform. With that one simple adjustment our success returned. I hate to imagine all the exciting discoveries we would have missed if we had oriented the speakers wrongly prior to our first experiment.

Soon, we discovered that smaller speakers that emitted sound from one location worked much better for our experiments. With their sensitive hearing, our bats could tell when the sound wasn't coming from a single source as it would from a real frog.

With that discovery came new innovations in our testing. In addition to presenting calls to individual captive bats in our flight cage, we could now carry pairs of pocket-sized recorders deep into the forest. Thus we could test larger numbers of bats without having to catch and tame them. We purchased microcassette Pearlcorder tape recorders and made our own 40-foot-long lead wires with switches for remote control. While one of us continued nightly lab tests in our flight cage, the other would hike island trails in search of good locations to test bats in the wild.

On Barro Colorado Island, nearly all the trails are in steep, hilly terrain, often slickened by sudden downpours. Carrying a scope, infrared light, and tripod, a lightweight blind, tape recorders, lead wires, a folding stool, and rain gear was arduous. Heavy rain showers were also problematic, though we could usually tell when these were approaching, thanks to the loud howls of troops of howler monkeys. Common on the island, as rain hit successive groups of howlers, their protests could be heard from afar.

One especially frustrating night, the howler monkeys gave

ample warning of rain as I scrambled to get my gear packed. But the shower turned into a deluge that caused my normally trusty electric headlamp to fail. I was a half-mile from the research station, left in pitch-black darkness. Using my folding umbrella was like trying to stop a river with a canoe paddle, and I had no spare light.

It could be hours before anyone came looking for me. Luckily, there were only a few trail intersections between me and the station, so I was able to painstakingly feel for the hard-packed trail surface with my feet and for denser vegetation along the sides with my hands. Inching my way along, hours later I spotted lights ahead just as colleagues were about to come looking for me.

Soon we realized we could solve a lot of problems by purchasing a small Boston Whaler and an outboard motor, allowing us to navigate the surrounding waters of Gatun Lake, part of the Panama Canal. That would make it much easier to reach distant locations around the island. But even that idea wasn't trouble free.

We couldn't run a motorboat in the canal without first obtaining a pilot's license. We had to pass rigorous exams on navigation, all in seaman's language. Then we had to learn the art of navigating a lake filled with barely submerged tree stumps. Blinking lights identified the main shipping channel, but nighttime navigation, even a few feet beyond, could be a real challenge. Most perilous were the four- to five-foot-tall, nearly vertical wakes left by tugboat captains who illegally sped under cover of darkness. Sometimes a tug would disappear around a small island without even being seen.

One night after completing night-vision experiments on the far side of the main island, I had successfully navigated the obstacle course back to the shipping canal and relaxed in what I thought of as relative safety. It was nearly midnight, and I was speeding along the canal unaware that a tugboat had just passed. The beauty of the starlit sky overhead had me nearly mesmerized. Then, without warning, I was suddenly thrown violently to the bottom of my boat. The boat and I were airborne for so long it seemed we'd never come down. When we did land, it was with a crash that would have shattered a lesser boat, and we were immediately airborne again. We had simply landed atop another wake and skipped ahead. Miraculously, our night-vision scope remained in the boat with me, and I soon arrived back at the research station unscathed.

Despite the hassles, most of the time I loved being out in the forest at night. It was fun to identify the owl-like hoots of small night monkeys and the whistling calls of kinkajous or to watch inch-long click beetles flying through the forest with their brightly glowing eyes that looked monstrous in my night-vision scope.

Of course my primary focus centered on the two small tape recorders. At dusk I'd find an opening through which I could watch them with a telephoto lens. I would place them exactly 40 feet in front of me and 12 feet apart. Then I would run all the same tests in the field that Mike was running in the flight cage – for example, providing a choice between a túngara frog and a poisonous toad, or the call of a túngara frog versus a South American bullfrog large enough to eat bats. Both túngara and bullfrogs produced similar whinelike calls except that the latter was much louder. The bats would demonstrate their choices by hovering over one speaker or the other as the recordings continuously played. Even when we adjusted volumes so the túngara and bullfrog calls were the same, we didn't fool the bats.

One night, I set up not far from the territory of a real bullfrog, who apparently took exception to the interloper. He called louder and louder, and when he couldn't scare off his presumed competitor, he came hopping over to my speaker. He stopped about three feet away and gave an extra forceful call. The speaker didn't quit or move away, so the frustrated frog attacked, jumping right on top of it. When the speaker couldn't be stopped, he fled and was not seen or heard from again.

Later the same evening, a philander opossum was attracted to my túngara frog call. This opossum is about the size of a large rat. It is commonly called the four-eyed opossum because it has a bright, cream-colored spot above each eye. It is frequently carnivorous, and in subsequent tests we demonstrated that, like our bats, it homes in on frog calls. My recorded calls even attracted hungry land crabs.

An ocelot provided one of my greatest thrills. On a slow night with few bat visitors at my speakers, I suddenly noticed two brilliant eyes approaching. The reflected eyeshine was so bright that at first I thought it might be a puma. But as it came closer, I could see by its smaller size, striped face and neck, and spotted body that it was an ocelot. Originally, it seemed attracted by my frog calls. Then it noticed the green light from my night-vision scope dimly illuminating my face. It calmly walked right up to investigate. I held my breath and froze to avoid frightening this beautiful creature. To my amazement, we ended up staring at each other from opposite sides of my mosquito net blind, so close I could have reached out and touched it. Though bats did eventually come to my speakers that evening, they were anticlimactic.

Mike had become an immediate and invaluable collaborator and friend. With his enthusiastic participation and that of our tireless field assistants, we often could run multiple experiments in the same night. Nevertheless, understanding our bats was crucial, something each of us had to learn. Initially, not fully understanding the intelligence of our bats was a serious limiting factor.

It also helped when we learned that, like people, each had its own unique intelligence and personality.

Training newly caught bats was essential to overcoming their fear sufficiently to allow testing in our flight cage lab. Successful training required patience – learning to read the animal's body language and understanding the importance of providing sufficient food, but not enough to eliminate their incentive to hunt. It was a delicate balance, not easily mastered. We learned that around the time of a full moon, frogs could spot bats approaching and take evasive measures, so bats caught at that time were low on energy and required more food to achieve the same performance. Everything hinged on providing or withholding rewards.

The first time I assigned assistant Steve Kern the task of training a frog-eating bat to come to our hands for food, he was a miserable failure in a most unexpected manner. I had emphasized the importance of "making friends," and he took me quite seriously. When we arrived the next evening to test the bat's choices, our newly tamed bat wouldn't perform. It only wanted to come to us for food. And when we didn't hold out our hands as expected, it simply landed on our shoulders and waited. It was frustrating, but I couldn't help but laugh.

Later that night, we had to feed and release the bat untested. It was just too tame. When it flew off into the forest, I assumed that was the last we'd see of it. Steve and I left the forest where we had released it and walked back to my living quarters. We were discussing the next day's plans, standing beneath a bright outdoor light. As I gestured in conversation, a large bat attempted to land on my hand. I was so taken by surprise that I actually wondered if I was experiencing my first bat attack. Steve exclaimed, "It's just our bat looking for another handout." That particular bat was easily recognized, because it had lost half of one of its large ears. Over the next several minutes it returned three more times for small minnows, and we were happy to oblige.

More than 20 years would pass before bat biologist Rachel Page would decide to further investigate frog-eating bat intelligence and memory. First, she documented conclusively that these bats do not need to inherit knowledge of frog calls. They do inherit remarkable learning and memory capabilities. In controlled flight cage experiments, she discovered that in as few as five trials, frog-eating bats could be taught to reverse their call preferences, responding to the calls of poisonous toads instead of túngara frogs. And in just a few more trials that knowledge too could be reversed.

What astounded even Mike and me was her discovery that tagged individuals that had been briefly trained and released retained their knowledge when inadvertently recaptured one to two years later. In her flight cage, they still flew to her hand on call without further training. As of this writing, Page has six examples of such memory.

My favorite observations continued to be through the night-vision scope out in the forest, and I never had more fun than at Weir Pond. Following my first night of discovery there, I would return often, armed with my scope and a microcassette tape recorder for recording rapid sequences of observations. Every night was different. Sometimes, when fewer than 50 túngara frogs were present, they were extremely careful not to call too often, too loud, or with added chuck calls, while on other nights more than 400 assembled and called with great abandon, ignoring the threat of bats. I would also notice differences due to moonlight.

In just five nights, I would observe 95 capture attempts, mostly successful. I never tired of watching or of developing new hypotheses to test. When Cindy Taft transcribed my tape recordings into quantifiable field notes, she would tease me, "You sound like a sports announcer at a big game." It was that exciting.

It soon became clear that frogs who didn't call, or who stopped before a bat came too close, did not get caught. One mystery especially intrigued me. Why did whole frog choruses sometimes shut down almost instantly at the approach of a bat that would be ignored at other times? Could frogs actually detect approaching bats, and, if so, how could an alarm be communicated without imperiling the frog giving the warning?

To answer these questions, I returned to Weir Pond with my night-vision scope and tape recorder, counted numbers of túngara frogs present, and recorded the weather and moon conditions as well as light levels with a light meter. Then I spoke into my recorder whenever I saw a frog-eating or insect-eating bat reach the pond's edge. Since the frog chorus was nearly deafeningly loud, it was easy to tell from the recording whether or not they stopped calling. By repeating this procedure on nights with few versus many frogs present, and on dark versus moonlit nights, one thing became clear.

On all but the darkest nights, the frogs could detect an approaching frog-eating bat. If fewer than a hundred frogs were calling, the entire chorus would typically stop within less than a second. However, the frogs didn't stop on moonless, overcast nights, when it was impossible to see. They also tended to ignore bats, even on moonlit nights, if enough frogs joined the chorus to tip the odds in favor of playing the frog version of "Russian roulette" that I had observed on my first evening.

Serious questions remained. Wouldn't it be suicidal to give an alarm call, as seen in small birds? Predators normally catch birds only by surprise, and in the light of day they frequently are visible while still at a safe distance. Warning one's neighbors against their being surprised incurs minimal risk. How could frogs communicate the risk of an approaching bat so quickly without attracting an immediate attack? After all, a fast-moving, low-flying bat probably wasn't visible from more than two or three feet away.

To answer such questions, we needed more rigorous control. For this, we made tracings of frog-eating bat wings, as well as of smaller insect-eating species that also flew over frog ponds. We next made cardboard cutouts darkened to resemble real bats, and coated them in paraffin to be waterproof. Each was equipped with lead weights below and rollers on top so it could slide down a sloping fish line strung over a second breeding site, the rectangular Kodak Pond. The slope of the fish line allowed our models to move across the pond at about the same speed as flying bats.

For our first experiment, we chose a moonless, overcast night. We had rigged a rheostat-controlled light over the pond, so we could produce the equivalent of dim moonlight for a test run of a model bat over the pond. For the next, we turned the light off, alternating trials at well-spaced intervals.

The results were clear. The frogs responded to our models just as I'd seen them do at Weir Pond with real bats. The entire chorus ceased in less than a second in even faint light, but not without light. Our results also showed that the frogs could discriminate between a model frog-eating bat and a smaller insect-eater. They were spotting silhouettes against bits of sky seen through the forest canopy. This accounted for why frog-eating bats normally flew so low. They nearly skimmed the ground, like military planes flying low to evade enemy radar detection.

Last, I performed what for me was the most interesting experiment of all. I placed a Pearlcorder with a recording of one túngara frog's voice at the pond's edge, and with one of our remote switches, used it to join the túngara frog chorus. By then I believed there could be just one way to communicate a hunting bat's arrival without a frog essentially committing suicide.

What would happen if one frog suddenly quit calling in mid-call? I waited for the chorus of more than 50 frogs to resume. Then I switched off the recorder in mid-call. Instantly, the entire chorus ceased. It was like magic! I tried again and again, each time silencing a whole pond of frogs in less than a second. I repeated the experiment back at Weir Pond with the same result.

What intrigues me to this day is how up to a hundred frogs can all be so constantly aware of each other as to quit calling within less than a second after the first frog sees a bat and abruptly stops. It's a really clever alarm system that doesn't put anyone at increased risk.

When possible, we also listened for other frog species. Soon we discovered that only a few very large or poisonous frogs could afford to ignore bats. These could call as loud, long, and conspicuously as they wanted. One night, we heard a tree frog making seemingly high-risk calls. We found that it was safely calling from the center of a spiny bromeliad plant. Each species seemed to have developed its own approach to predator avoidance, sometimes to the extent that female frogs were apparently forced to evolve special hearing.

Pug-nosed tree frogs have multifaceted strategies that are especially fascinating. Like túngara frogs, males use both single-note and multi-note calls. But in our observations, unlike túngara frogs, males of this species never joined together in large choruses. In fact, they seemed to be territorial, never allowing another to call too close. Watching through my night-vision scope one night, I noticed one chase away a rival. Then I tried to record this species and failed to find a single one calling where I could tape it without loud background interference. I searched streambeds for a whole evening without finding a single one calling far enough from a waterfall or noisy riffle.

I was feeling discouraged after nearly a half-mile of futile search, when it finally dawned on me that this might not be mere coincidence. I knew from a previous discussion with bat ultrasound expert Jim Simmons that waterfalls produced intense, random sounds of a wide range. He had suggested that such sounds might jam bat echolocation or simply hurt their ears, as in the human response to fingernails scraping on a blackboard. Could it be that pug-nosed tree frogs were actually guarding waterfalls as courtship territories, because bats avoided these obnoxious sound combinations?

By the next evening, I had figured out that with a little practice, I could passably mimic the single-note call of a pug-nosed tree frog. It actually worked so well that I could use my fake calls to census the distribution, even of males that weren't actively calling. I would make a sharp *konk!* sound, and if a male was near, it would respond. I was surprised to find that they would come hopping right up to me to chase the interloper away.

One night, I was having fun watching an upset male hop up to chase the presumed intruder away when I was amazed to see a frog-eating bat catch him and carry him away; he was taken right out of my headlamp beam, only an arm's length away. From then on, I was a bit more careful with my newfound knowledge. I had inadvertently lured this frog into calling too frequently too far from his waterfall.

Paying closer attention to these frogs, I noticed that they liked to call in moderately bright light – for example, at dusk. As darkness descended, they first called less frequently, then quit using complex calls. In full darkness, they quit and climbed back up into the trees. Unlike the túngara frogs, they refused to call in the dark. Having observed a careless one caught right in front of my face, I was well aware of the threat from bats.

Later that night, I met with Mike at our flight cage. He had just given up on further testing of my waterfall hypothesis. He had finally gotten a usable pug-nosed tree frog recording and was testing a frog-eating bat's response to it when played alone versus next to a waterfall recording I had made on one of our little Pearlcorders. He had been quite surprised and reported that this bat actually showed a preference for the frog call when it was played near the waterfall sound.

By that time, I had used my "konking" technique to count numbers of frogs near versus away from waterfall noise and knew that they were clustering in the vicinity of such sounds. It made no sense. Would any species guard territories where bats preferred to hunt? Then Mike, the more technologically sophisticated of the two of us, wondered out loud, "Might it make a difference if we used my Nagra tape recorder to record a waterfall? That would include a far superior range of frequencies." Not ready to give up, I urged him to try.

The next day, Mike made the improved recording, and we designed an experiment to see if pug-nosed tree frogs could be lured into calling again after dark if I used my rheostat-controlled headlamp with a diffuser lens to imitate moonlight.

The night before, Steve Kern had caught a new bat, which he was now able to expertly train. We had learned to lure frog-eating bats into our mist nets by playing túngara frog calls under a bush with a net in front. Frog-eating bats are experts at avoiding nets. But by luring them to search in vegetation, where nets are hard to detect, we caught them easily, often within a minute of starting a call recording.

Now Mike could test his new waterfall tape. To his delight,

when faced with a choice involving the improved waterfall recording, the bat significantly preferred the call located farthest from the waterfall sound, explaining at least one good reason for the frogs' love of waterfalls.

Simultaneously, I was out testing the impact of light. I had already noticed that our frogs, though territorial, would permit additional frogs to call around the same waterfall as long as they were well spaced out. I had brought my folding stool and special headlamp and sat near such a location. The amorous frogs were calling lustily at dusk. But as darkness closed in, calling gradually ceased, and the frogs began climbing back into the trees. I had already mounted my light strategically on a nearby tripod, and now I gradually illuminated the area to approximate the brightness of moonlight.

Sure enough, the frogs reversed course, resumed their former positions, and began calling again. In the now brighter light, they even dared use their complex, riskier calls. For the next several nights I repeated this experiment, and the results were consistent. Pug-nosed tree frogs wouldn't call without at least dim light.

To determine if their preference for calling only at dusk or in moonlight was for the purpose of watching for bats, we again brought out our model frog-eating bat. The response was less spectacular than when we tried with túngara frogs, but these frogs did respond. For at least five minutes after the model passed over, frogs reduced their call rates, used simpler calls, and attempted to call from more concealed locations. A common response involved backing under a large fallen leaf, where the frog couldn't be surprised from behind.

We had earlier recognized that pug-nosed tree frogs also synchronized their calls with one another. This made it difficult for me to locate calling individuals and sometimes was so cleverly done that I couldn't be sure how many were calling. When I was "konking" to find individuals, I often had to repeat myself to be sure a frog had answered. That couldn't help female frogs. But could it hinder hunting bats?

To test this hypothesis, Mike again played call recordings in our flight cage lab. Not surprisingly, frog-eating bats preferred asynchronous to synchronous calls, and also chose complex over simple.

My partnership with Mike resulted in our publishing 13 scientific papers together, the first a cover story in the journal *Science*, immediately followed by an article in the January 1982 issue of *National Geographic*. We eventually documented that frog-eating bats are major predators of a wide variety of small frogs, and that they indeed identify frog calls so cleverly that, even when we later altered recorded calls in an effort to trick them, they weren't fooled.

We also discovered that frog-eating bats have evolved unique hearing that enables them to hear low-frequency frog calls in addition to their own high-frequency echolocation calls, an ability previously unknown in mammals. Clearly these bats have had a significant impact on the evolution of frog courtship behavior.

Experiments on frog-eating bat learning and memory have already demonstrated abilities that far exceed anything we could have predicted, and there is much more to be discovered. The behavior, intelligence, and learning capabilities of most of the world's more than 1,300 bat species have yet to be investigated, and my personal experience suggests that frog-eating bats may face stiff intelligence competition from other species. The doors remain wide open for the next generation of young scientists to make exciting discoveries.

CHAPTER 7

FINDING AMERICA'S MOST ELUSIVE BATS

"STEP ASIDE while we inspect your plane! Where are you from?" We were being interrogated by two U.S. Drug Enforcement agents, backed up by the Texas Rangers and a local sheriff. Unbeknownst to us, we had been mistaken for drug traffickers when our small plane was spotted flying low near the Mexico border. Two enforcement planes had been tailing us and had radioed ahead for reinforcements. When we landed in the small Hondo, Texas, airport to refuel, the agents were hoping for a big drug bust, and they were unhappy to learn they had wasted time chasing bat biologists.

Our pilot, Don Grantges, was an early trustee of Bat Conservation International who enjoyed flying his 12-year-old son, Bert, and me down deep canyons of the Big Bend area of West Texas in search of isolated water holes where we might be able to net one of America's rarest and most spectacular mammals, the elusive spotted bat (*Euderma maculatum*).

This species was on my Most Wanted list so that I could show off the beauty of American bats. It has long, soft fur; snow-white underparts; and a jet-black back with three large white spots, one on each shoulder and one on the rump. The wings and huge ears are translucent pink.

Spotted bats were known throughout most of western North America, from Mexico to Canada, but sightings were rare, even by bat researchers. We would brave 120-degree summer heat in Big Bend, Texas, and frigid winter weather near St. George, Utah, in order to finally photograph them successfully.

The hunt began in May 1987 in the Big Bend National Park. Don Grantges was a former Navy aircraft carrier pilot who loved challenges. His flying antics in his Beechcraft Bonanza, zooming low down cliff-walled canyons, had attracted more attention than we wanted. When we had seemingly popped up out of nowhere, drug enforcement agents were immediately suspicious. Low-flying planes along the Mexico border were nothing but trouble, so they were rather surprised to encounter such a plane whose occupants claimed to be looking for bats.

Don helped on weekends, then flew home to run his Fort Worth real estate business during the week. His wife, Carol, and son Bert put in the all-nighters helping net for spotted bats during the week. We had chosen the month of May because that was the hottest, driest time, the period in which bats were most attracted to remote water holes.

We had not yet learned that spotted bats drink only from large ponds, normally at least 100 feet long. Because their wings are not extra-long and narrow, like those of free-tailed bats, no one initially suspected they would require large areas of open water to drink, and, frankly, I still don't understand it. In retrospect, most of the sites we visited in the park were too small for spotted bats, though more than a dozen other species could use them. The Chisos Basin, the park's lowest area, also contained some of its most isolated water holes.

Simple survival in this hottest part of the park was a challenge. Struggling against 120-degree heat, Bert and I rounded one last bend in the Ernst Tinaja Canyon, hoping to find a remote watering hole we might have missed seeing from his father's plane. Ahead, for as far as we could see, stretched nothing but more hot sand and shimmering mirages. Discouraged, and at risk of succumbing to heat prostration, we turned back.

Finally, late in the afternoon, while searching one more canyon, we spotted a series of three water-filled rock cavities known as *tinajas*. Carved into solid bedrock over many thousands of years, these provide the area's best natural water-storage tanks. Three pools, the largest 30 feet long, provided what appeared to be an ideal location for finding thirsty bats. They were surrounded by towering rock walls, more than ten miles from the nearest alternative water, making this the equivalent of Grand Central Station for thirsty wildlife.

We quickly set a net over each pool, immersed in the scent of creosote, lechuguilla, and cactus plants. By now we'd been joined by Bert's mother, Carol, who helped organize our equipment. She had brought an ice chest with food and drinks, a tarp, and sleeping bags, so we could take turns resting during the night. Everything ready, Bert and Carol waited for sundown while I climbed an adjacent ridge to take a few pictures, hoping to document the site of our first spotted bat capture.

As I returned, only about 30 feet above them, I inadvertently kicked a bit of gravel into an eroded cut in the otherwise solid rock. It apparently landed on an unexpected visitor. The largest mountain lion I have ever seen sprang straight up from the depression, landing less than ten feet in front of me. If looks could kill, I'd have died on the spot. That cat gave me the dirtiest look I've ever seen, then just stood, glaring.

All I could think of was what a rare moment of beauty this was. Wanting to share it, I yelled, "Bert! Carol! Come look. There's a mountain lion standing right here in front of me." I suppose I shouldn't have been surprised when neither showed up. The cat twitched its long tail several times, then sauntered off. I couldn't fault the malicious stare. After all, I had destroyed its chance of a javelina dinner.

I was beside myself with frustration that Carol and Bert had missed a once-in-a-lifetime opportunity, so Carol suggested we climb up the ridge to see if we might still see it. Sure enough, less than 70 feet beyond the crest, our cat was stretched out on a large rock, casually watching the sunset.

Soon after sundown, we encountered an unanticipated problem. Our nets were swamped with hundreds of individuals of common bat species, precluding any likelihood of capturing spotted bats even if they had been present. And even if there had been time for rest, Carol and Bert were disinclined to retreat to the sleeping area with a hungry mountain lion prowling nearby, not to mention having discovered a scorpion and a rattlesnake within a few feet of their sleeping bags.

At least this was a great opportunity to teach Bert and Carol about desert bats. Even before sundown we could see canyon bats (*Parastrellus hesperus*) flying high overhead. With bodies no larger than the last digit of my little finger, these are one of North America's smallest bats. When one finally swooped low to drink and was caught, Bert shouted, "Got one!" and ran to get it. His mother came to see and exclaimed, "It's cute!" This bat's large, black eyes were hard to resist. Its body fur was a beautiful golden tan except for an inky black facemask and ears. I explained that this species lives in narrow, cliff-face crevices that must be safe from snakes and other predators and oriented just right to the sun so as to gain sufficient warmth without overheating. Black faces make these bats nearly invisible as they peer out from dark crevices.

Almost an hour later, we began catching pallid bats (*Antrozous pallidus*), lots of them. Most were covered with yellow pollen. We assumed that they acquired the pollen only as they chased large sphinx moths that hovered at flowers. Twenty years later, however, biologists would document these bats to be deliberate nectar-feeders in the Sonoran Desert where they effectively pollinate giant cacti. Pallid bats are famous for their predilection for feeding on scorpions and giant centipedes, unharmed by their venomous stings. They rely on large ears to listen for faint footsteps, typically plucking prey directly from the ground.

Soon we were removing a steady stream of thirsty bats. Then I heard Bert exclaim, "Wow! What kind is this? It's unbelievable."

At a glance I could see it was a ghost-faced bat (*Mormoops megalophylla*). This bat is unmistakable. Like some dogs, it's so strange, it's endearing. The eyes are located in its ears, hence its name. This was a solitary male. Females of the species form large nursery colonies in caves. Farther south, such colonies may contain hundreds of thousands of individuals.

The Big Bend National Park benefited from our failed searches. We provided them with photos of most bat species of the area, and those are relied on to this day as a basis for popular public education programs.

Our next search was near St. George, Utah, where a graduate school friend had netted spotted bats over a desert spring. He had found these bats in late winter, when most other bats were inactive.

Don Grantges flew Bert and me to the nearest airport, where we rented a four-wheel-drive vehicle, met Bat Conservation

International volunteer Dee Roper Lockwood, and drove to a nearby motel. The manager looked at me as if I might be deranged when I asked him to remove the furniture from one room so I could set up my photo studio for bats. But as I had done in Africa, I showed him my most recent *National Geographic* article, and he quickly warmed to the idea. He said, "Just let me know whatever you need. We could probably even tear out a wall to give you more work space."

Later, as we watched snowflakes fall, I wondered if I might indeed be crazy. After all our previous failures, it seemed a bit presumptuous to rent a special room and set up a photo studio for spotted bats that had yet to be captured. The weather didn't seem promising. The next day the temperature rose only to the mid-40s, not to mention a 30-mile-per-hour wind. Nonetheless, despite continuing wind and a hard freeze, that night Bert, Dee, and I set nine nets across a series of 30- to 150-foot-long pools. By midnight, we still had not seen one bat. Chilled to the bone, we closed our nets and retreated to the warmth of the motel.

The next night, under similar conditions, but minus the wind, we tried again. The nighttime temperature was still well below freezing, but at nearly midnight, we finally caught our first spotted bat. Jubilant, we rushed to the net. We sobered quickly, though, when we saw that the bat had been killed moments before by an owl. What miserable luck!

By the next day, the temperature rose to 55 degrees, but we again faced winds gusting to 40 miles per hour. Hoping it would calm by sundown, and now running short of time, we decided to put in an all-nighter. Equipped with what we believed would be warm enough clothes, two large tarps, and our sleeping bags, we returned.

Unfortunately, the winds didn't abate, and the temperature

plummeted soon after sundown. We took turns getting up to check the nets every half-hour. Chilled to the bone, we were far too cold to sleep. In fact, my notes from the next morning state, "We about froze to death." Our sleeping bags and tarps were covered in ice condensed from our breath. Of course the ponds froze over, so we weren't surprised not to catch any bats.

The following day brought warmer temperatures and a 40-degree night in which we finally caught three spotted bats. We returned to our studio with high hopes, only to find that two of our three bats were in no mood to cooperate. Only one would eat mealworms in captivity, meaning the others had to be returned to the capture site and released.

Fortunately, the one willing to eat in captivity had an ideal temperament. His appetite was endless, and he'd do almost anything for food. Using natural rock materials from local cliff faces, I was able to create an artificial set. I'd release our bat, nicknamed Porky, into one of the crevices, and he quickly learned to pop out to look for food when called. His happy pictures were no accident!

To this day, little is known about these mysterious bats. Why do they have such striking coloration, huge ears, and patchy distributions?

What we do know is fascinating. Spotted bats appear to feed almost exclusively on moths whose ears, known as tympana, are tuned to hear bat echolocation frequencies between 20 and 60 kilohertz, the dominant hunting frequencies used by most bats.

Echolocation expert Brock Fenton and associates discovered that by using hunting cries whose peak energy was concentrated around 10 kilohertz, spotted bats achieve two goals. The bats prevent listening moths from detecting their approach until it's too late to take evasive action, such as diving into foliage. And their low-frequency calls make it easier for spotted bats to hear each other at greater distances and avoid encroaching on one another's feeding territories.

This approach is well designed for use in open areas where low-frequency calls can detect relatively large prey at greater distances. It also may explain why spotted bats are rare among bats and never form large colonies or feed in forest vegetation like many other species. Their low-frequency calls don't work well in cluttered environments, and their stealth strategy can succeed only as long as spotted bats are a small, well-spaced minority among bats. That way, moths can't afford to retune their hearing.

Illustrative of moth ability to adapt, in Hawaii, where the hoary bat (*Lasiurus cinereus*) is the only bat species present, moths have retuned their hearing to concentrate on this bat's lower-frequency social calls instead of listening for its echolocation signals. By focusing on the frequently emitted social calls, they can detect approaching hoary bats from longer distances, allowing more time to take evasive action. Continental moths cannot afford to retune, because they would then be too vulnerable to other bats. Clearly, bats and insects are locked in a constant arms race.

Some bats, such as horseshoe bats (Rhinolophidae) of the Old World, rely on an opposite strategy. They make their calls undetectable by moths and other listening insects by using ultra-high frequencies, above 100 kilohertz. Such an approach is far better suited to fine resolution close-up, as required for hunting in vegetation.

After finally catching and photographing North America's most spectacular bat, my next goal was to find the largest bat of the United States, the western bonneted bat (*Eumops perotis*). This giant among North American species weighs two and a half ounces (five times the weight of a spotted bat) and has a nearly two-foot wingspan. It lives in widely scattered locations, mostly arid, from southwestern Canada south to Peru. Like the spotted bat, it is rarely caught in bat researchers' nets so is little known. It takes echolocation to a new low, relying on the lowest frequencies yet known for a bat, below 10 kilohertz. Bonneted bat echolocation is so poorly studied that we can only speculate on their hunting strategies. Terry Vaughan, one of the first bat biologists to study western bonneted bats, told me he sometimes could hear them from a thousand feet away.

Despite their distinctively low-frequency calls, western bonneted bat echolocation remains unstudied. Like spotted bats, they can detect obstacles and large insects at much greater distances than most bats, but with minimal detailed information. For the detail needed during pursuit and capture, they may rely on sounds produced by flying insects.

Entering the Fresno Canyon area of what is now part of the Big Bend Ranch State Park, southeast of Presidio, Texas, my volunteer field assistant, Sally Smyth, and I hoped to find a pool of water large enough to entice bonneted bats to drink. We finally breathed a big sigh of relief. Ahead, the canyon widened out to a gravelly, sandy, dry riverbed that was easily drivable in my Toyota 4Runner.

We had just driven through ten miles of nearly impassable terrain. An old ranch road had long ago been washed away by floods, leaving a rugged riverbed as the only access. We had lost count of all the boulders we'd moved or built around so we could pass over the top, not to mention the steep banks we'd had to grade to get up or down. Our only tools were a pick and a small shovel.

A colleague had told me of a remote pool where bat researchers, some 20 years earlier, had caught bonneted bats. Unaware that the entry road no longer existed, we'd set out on a more-than-25-mile drive from the last highway. Had we understood the ruggedness of the canyons we'd have to traverse without roads, we likely wouldn't have started. Even worse, after fighting our way through a grueling ten-mile stretch of boulders and vertical banks, we found that no water remained where the bats had originally been found.

Now discouraged, but exhilarated by the rugged beauty of this remote stretch of Chihuahuan Desert, and still hopeful of finding another suitable pool, we accelerated to the heady speed of 20 miles an hour. Then, suddenly, we stopped so abruptly it was like we'd struck a rock wall. Barely even able to open the doors, we were incredulous to discover ourselves mired in quicksand. Yes, quicksand in a supposedly dry desert.

With no chance of finding help, we assessed a seemingly hopeless situation. All four wheels were mired, with both axles submerged in a pool of fine, water-permeated silt. The surface looked just like the rest of the dry riverbed, but just beneath lay a nearly 3-foot-deep by 30-foot-long pool of quicksand. Using a small camp shovel, I began digging beneath the rear frame, the area of least problem. For every two shovelfuls I removed, at least one surged back. However, we found a ready supply of large, flat rocks nearby and quickly developed a system.

After digging like mad for several minutes, I'd ram a large, flat rock into the cavity. Then we'd wedge our jack into the remaining space below the frame and jack the vehicle up until another rock could be positioned beneath the nearest wheel. After placing one or more flat rocks beneath a wheel, we'd lower the jack and place additional rocks beneath it. By alternating lifting and placing new rocks, we gradually lifted all four wheels a little more than two feet. We also built two flat-rock tracks to back out on.

The work was excruciating in the 95-degree heat, but approximately four hours later, with a lot of backbreaking assistance from Sally, we were on our way again, this time carefully avoiding low areas in the riverbed. About two miles farther on, we took a turn that looked promising, drove as far as we could, and hiked another half mile, periodically finding water, but not enough to attract bonneted bats.

These bats have narrow wings, adapted for extraordinarily fast, long-distance flight. Like miniature jets, they have low maneuverability, in sharp contrast to the frog-eating bats I'd studied in Panama. The latter, at the opposite end of the bat flight spectrum, have short, broad wings with which they can lift themselves straight up or even go in reverse while carrying a frog.

Bonneted bats, because of their high-speed flight, must find a pool at least 100 feet long before they can swoop to drink. Their specialized flight enables them to cover long distances between patches of insect abundance in unpredictable desert habitats, but greatly limits their ability to find a drink.

The odds of our finding a pool of the required length in a narrow canyon seemed slim, but just as we were about to give up, we climbed one more rise in the streambed, and there, right before our eyes, was the equivalent of a bat man's pot of gold: a sparkling clear pool of water, 120 feet long, 12 feet wide, and just 2 feet deep. We hoped that the unobstructed length would be sufficient for bonneted bats to drink in flight. The narrow width and shallow depth would make bat netting far easier than anticipated.

Hardly able to contain our excitement, we ran back to the vehicle to get our headlamps, nets, poles, cloth bat bags, and a night-vision scope. By sundown, we had a net set over the middle of the pond and could hardly wait for bats to arrive. Unfortunately, by area standards, a 95-degree day was a cool one, apparently insufficiently hot to force bonneted bats to drink. Late that evening, greatly disappointed, we packed our equipment and began the tortuous drive back to the motel, fervently hoping not to get stuck again. We made it, but not until the wee hours. My vehicle had sustained some $3,000 worth of damage, but that was the least of my worries.

The next day, the forecast was for warmer weather, with a high of 105 degrees. Sally looked like she thought I'd lost my mind when I decided to return on the chance that higher temperatures would drive bonneted bats to drink. At least this time we didn't have to build a road, and we didn't get stuck.

Soon after dark we heard the first unmistakable bonneted bats. They were feeding high up along the canyon walls. The suspense was excruciating as we waited to see if they would come down to drink. By nine o'clock, we were enthralled as we listened to a cacophony of piercing calls, periodically punctuated by machine gun–like bursts, known as feeding buzzes, during the final pursuit of prey.

The first bonneted bat to approach our pond was easy to identify. Its staccato calls could not be mistaken as it made a power dive just over the net. Moments later, it circled back, and I watched, riveted to my night-vision scope as it approached, this time on a collision course. The bat struck so hard, it was like catching a small duck. It stretched the net several feet and was thrown back, similar to a person jumping on a trampoline. Unharmed, it fell free and swam rapidly for shore as I dashed to intercept.

It was truly a gentle giant. Despite its unhappiness at being caught, it made no attempt to bite. With its strange face and huge ears, it looked like something left over from the age of dinosaurs. To this day, the western bonneted bat ranks among my all-time favorite animals.

By 9:30 we had caught three more and estimated that at least 25 were circling overhead. The sight and sounds were magical. But as quickly as possible, we removed our net to avoid unnecessarily disturbing these never-to-be-forgotten creatures. The trip back to the motel where we had set up my photo studio seemed a whole lot shorter, though in reality it took most of the night. By that time our bats were quite hungry and readily ate mealworms from our hands. Their calm personalities reminded me of the Saint Bernard dogs I had grown up with as a teenager. Though we were able to get nice pictures, we could not show these bats in flight. Such speedsters simply cannot fly in a confined area.

My next quarry was very different. The Keen's myotis (*Myotis keenii*) is a small bat, weighing less than two-tenths of an ounce, about a tenth as much as a western bonneted bat. Like many members of its genus, it is brown and unremarkable except for being rare and virtually unknown. Bat biologists can't even reliably identify this bat except through detailed skull measurements or genetic analysis. It is known only in a few locations, from British Columbia in Canada and from Alaska and Washington in the United States, where it is thought to require old-growth forests.

A small nursery colony of these bats was found in a unique rock crevice on Hot Springs Island, one of more than 150 islands that comprise the Queen Charlotte Archipelago. This crevice provided an ideal home. About three-quarters of an inch wide, and two to three feet deep, it was heated by hot spring water that trickled out along the bottom. By adjusting their position between the hot water and cool outside air, these bats could always find ideal, incubator-like conditions for rearing their young.

Hot Springs Island, little more than 100 yards in diameter, is located approximately 75 miles south of Alaska and 60 miles west of British Columbia. In such a remote location, one might reasonably expect to be alone. Nevertheless, just as I was ready to take my first photograph of the island's unique bat roost, I caught movement in my peripheral vision and glanced up to see a rapidly retreating nude woman. I had thought that my assistant and I were the only people on the island.

This was just one more surprise in a day of unexpected happenings. We had come to this remote island, funded by *National Geographic* magazine, with special permission from the island's Haida Indian owners. We expected to spend five days in isolation photographing the colony of Keen's myotis that lived just above the beach.

Our first big surprise had come several hours earlier when our chartered pontoon plane arrived at low tide. Instead of pulling up to a sandy place on the beach as expected, we faced sharp rocks some 50 feet from shore. The pilot steadfastly refused to risk his plane by going closer, so I had to jump into waist-deep 40-degree water and wade ashore. Fortunately, I found a small boat at the Haida hunting cabin where we'd be working. By the time we'd ferried 600 pounds of food and gear from the plane to the shore, I was chilled to the bone. It was August 3, but even the air temperature hovered in the low 40s.

Marveling at the ancient rainforest trees draped in a profusion of mosses and lichens, with majestic snow-capped peaks just across the bay in front, the last thing we had thought of was meeting other humans, much less people skinny-dipping in the hot springs. Within an hour, we would need to set a 42-foot mist net right above the location from which we now clearly heard voices. We knew overnight camping wasn't allowed on the island, but our neighbors didn't seem to be leaving, and we needed to get ready to catch bats.

Finally, we gathered our gear and headed slowly down the trail leading to the main hot springs, being sure to make plenty of noise so they'd know we were coming. We didn't want to surprise anyone, but as it turned out, it didn't make any difference. When we finally rounded the last bend, we found a half-dozen people quite comfortable with their nudity, which we politely tried to ignore.

I explained that we were there to photograph bats and hoped they didn't mind our intrusion. On the contrary, they were completely fascinated to meet bat researchers. One of them promptly asked, "Do you happen to know a bat researcher named Merlin Tuttle?"

I was a bit surprised, of course, but said, "Yes, I am Merlin Tuttle."

It turned out that they had just arrived via a large yacht carrying a pontoon plane. They were parked just out of sight around the tip of the island, and they had been reading Diane Ackerman's latest book, *The Moon by Whale Light*, which contained a chapter about me and bats. Now they enthusiastically volunteered to help, and their assistance turned out to be essential. Since the bats were far too shy to be photographed in their rock crevices below the hot springs, we had to again set up my studio and assemble rock crevices indistinguishable from the bats' natural home. The bats would be trained to emerge on call for mealworm rewards, this time in a small Indian cabin.

The bats could be caught only by netting them as they emerged at dusk, and we soon discovered that this required teamwork assistance. The bats had to pass over a small ridge about 40 feet from the hot springs pool, but they could easily detect and avoid nets set on the ridgetop. To catch them, we needed more hands and eyes. First someone in the pool had to warn us that a bat was coming. Then two people holding the net poles had to suddenly swing our net up into the approaching bat's face. Only in that manner could we catch them, and I had to be free to grab and disentangle bats as they were caught. I hate to think how we might have fared without help from our new acquaintances.

I would eventually go on to photograph all 46 United States bat species, thanks to the deeply appreciated help of many, an essential first step in conserving America's bats.

CHAPTER 8

CACTI THAT COMPETE FOR BATS

A SUDDEN HISS froze me in my tracks. Anticipating having cornered a jaguar, I flattened myself against the nearest cave wall, hoping to allow it an avenue of escape. I had entered a six-foot-wide passage in Cueva del Tigre in northwestern Mexico, looking for lesser long-nosed bats (*Leptonycteris yerbabuenae*). Another loud hiss, and I was slowly backing out, heart in my throat. By the time I neared the entrance, it occurred to me to look at the dusty floor for tracks. When I saw only vulture tracks, I had a good laugh and cautiously worked my way back for a look.

Just as I'd suspected, I found two snowy white balls of down, each nearly a foot tall, now hissing like little steam engines. They were turkey vulture chicks, their loud vocalizations easily explaining how the cave got its name. In reality, they were probably doing a great job of protecting the bats from human disturbance.

I had come to Mexico to provide the first photographic documentation of lesser long-nosed bats pollinating organ pipe and saguaro cactus and visited Cueva del Tigre to catch several of these bats to use in my studio. I planned to work in a hotel room in Kino Bay, a small coastal town in the Sonoran Desert of western Mexico. A night later, the bats performed so well, in approaching flowering branches of saguaro and organ pipe cactus, that I several times risked overheating my flashes.

The next day, I noticed that flowers of the also abundant cardon cactus had opened only at night and were closing by the following morning. This giant is the world's largest cactus, sometimes towering 50 feet tall. I wondered if it might also be relying on bats for pollination.

That night, I brought cardon flowers to the studio, and the bats immediately went to them. Several evenings later, armed with my night-vision scope on a tripod, an infrared headlamp, a folding chair, a warm jacket, and a notebook, I set out to document bat pollination of cardon in the wild.

The cactus I selected was multibranched, each arm some 13 inches in diameter and 10 to 15 feet long. Two branches bore large buds that I assumed would open that evening. I didn't want a really tall plant for fear of not being able to see the flowers well enough.

During midday heat, the Sonoran Desert felt like an introduction to hell. As the sun set, however, it more closely resembled heaven on earth. The temperature plummeted. The sky lit up with beautiful pastel colors, and an amazing assortment of wildlife began to emerge. While looking for a comfortable vantage point, I met my first Gila monster. Its beady black eyes and face weren't exactly winsome, but the two-foot-long venomous lizard meant me no harm. Its bright pink and black body could have made a tattoo artist jealous. Sightings of this shy creature are rare, so I considered myself lucky. This lizard, unlike most others, lives almost exclusively in subterranean burrows.

Gila woodpeckers chattered noisily from their nearby nest in a cactus cavity, and a coyote family serenaded me from afar. A forest of columnar cacti, backlit by the setting sun, stood out like desert sentinels. Three species dominated the landscape, providing critical food and shelter to a whole ecosystem of wildlife, from bats, pack rats, and squirrels to woodpeckers, doves, and owls.

As dusk gave way to night, the cardon flowers, and those of a nearby organ pipe cactus, opened as if by magic. Those of cardon had several-inch-long stalks that resembled fur-covered alligator skins. The open flowers displayed 20 or more narrow, white petals arranged like spokes on a bicycle wheel. The funnel-shaped interiors were a yellowish cream color. Male flowers were particularly impressive thanks to their heavy pollen loads. Funnel-shaped entries led to nectar-filled cavities.

A quarter moon rose in a crystal-clear sky adorned with stars so bright it felt as if I could reach out and touch them. Distracted by a flash of reflected light in my night-vision scope, I noted I'd been checked out by a large sphinx moth. It proceeded to the nearest cardon flower, where it hovered for a second, probing for nectar with its several-inch-long proboscis.

Hours went by, and I didn't see a single additional visitor. Having checked cardon flowers on previous evenings, I was well aware that the copious nectar and pollen they were providing should be strategically aimed at attracting bats. Just one of these flowers could satisfy a moth's entire needs for the night, leaving no incentive to move from plant to plant as required for cross-pollination.

But where were the bats? As midnight came and went, I became concerned. When I had inquired if there were caves or abandoned mines where bats might be found, locals had responded, "*Sí, pero nos quemamos para matar murciélagos.*" Yes, but we burn them to kill bats. They had warned me that if a bat flew over and urinated on me, I'd go blind. I had checked the nearest abandoned mine, and indeed it had sheltered nectar-eating bats until it had been burned. Both the bats' distinctive yellow droppings and the remains of a fire were unmistakable confirmation.

At nearly one in the morning, I was preparing to end my vigil in defeat when I heard the familiar wing beats of approaching nectar bats. A long-nosed bat, head already dusted in luminous pollen from another blossom, circled my flowers and left. Soon it returned, swooping up from below, briefly thrusting its long snout deep into a flower. This bat was followed by three more, each visiting a different flower in quick succession. As they departed, the bats' heads looked like they'd been dunked in a bag of flour.

My limited evidence seemed strong. Lesser long-nosed bats were probably this plant's most effective pollinators, but they had recently been declared endangered. Just how important were bats, and were there enough left to pollinate these dominant cacti? My casual observations of buds, flowers, and fruits suggested that cardon, organ pipe, and saguaro cacti might be staggering their peak flowering and fruiting seasons to reduce competition for bats.

In search of answers, I called Ted Fleming, a leading researcher of plant-visiting bats in the Latin American tropics. I reported my initial observations and offered to collaborate with him in writing a grant proposal to the National Geographic Committee for Research and Exploration. And because of the importance of such research to conservation, I also offered to recruit Bat Conservation International volunteers.

We began a multiyear project that would consume the remainder of his career and generate important impetus for conserving long-nosed bats and the Sonoran Desert ecosystem. When our grant was funded, we ran an ad for volunteers in Bat Conservation International's member magazine, *BATS*, which included the following warning: "Requirements? An ability to get along with people in close quarters, work long hours on little sleep, endure (sometimes) extreme heat, be strong enough to carry ladders, and not be afraid of heights, hard work or the dark." Ten of the finest volunteers a scientist has ever met joined us in the field, making an otherwise impossible task both fun and feasible.

We conducted fieldwork from April through June in 1989 and again in 1990. Since the lesser long-nosed bat had declined to the point of being federally listed as endangered, we asked, "Is this bat's decline causing reduced seed production in the Sonoran Desert's dominant cacti?" We also aimed to measure the relative effectiveness of bats, birds, and insects as the cacti's pollinators.

To find answers, we carried out a variety of experiments. Using bridal veil netting, we covered flowers either at night or during the day to compare the effectiveness of daytime (birds and bees) versus nighttime (bats and moths) pollinators. We also left some flowers uncovered throughout an entire 24-hour cycle, serving as controls, and continuously bagged others to test for possible self-pollination. In our final sample, we cross-pollinated flowers by hand prior to covering them. We tagged them all, and we censused them weekly to determine the proportions of experimental flowers that set fruit.

The nighttime pollination comparisons began at dusk and ended just before dawn, the time at which daytime experiments began. We also studied the timing, quantity, and quality of nectar production for each species, using small syringes to extract nectar every two hours from bagged flowers.

Each comparison required one of us to climb 5 to 25 feet up

a ladder held steady by a partner, often in gusting winds. We repeated this three times a day at 15 to 20 flowers. With swaying plants, prickly spines, angry bees, occasional rattlesnakes, and the desert floor full of easily collapsed rodent burrows, the work was challenging.

Peg Horner, the project's chief research assistant, remembers, "We were up at three in the morning, climbing cacti in the dark, hauling heavy extension ladders around from place to place, racing the clock to take night experiments down and put up day experiments before dawn – that was how a typical day began for many of us. It seemed we were always either carrying or climbing a ladder."

Volunteers took turns at a variety of tasks. Two at a time would trade off watching numbered flowers through our night-vision scope from dusk till dawn. The scope was mounted on a sturdy tripod, and the observer sat in a reclining lawn chair, providing a better angle for viewing activities at flowers high above. Participants meticulously recorded each pollinator visit while partners slept in the nearby vehicle. They changed places every two hours. At dawn, new volunteers took over, watching the same flowers with binoculars until they closed during the day.

Still other volunteers were assigned to radio-tracking duty. They worked with Ted to catch long-nosed bats in mist nets and use surgical glue to attach tiny radio transmitters to their backs. Bats carried these transmitters until they fell off 10 to 14 days later. By climbing to the tops of scattered hills, carrying receivers and antennae, volunteers could track the bats' nightly travels to feed, providing invaluable information on their effectiveness as long-distance pollinators and seed dispersers. When pollinated flowers produced fruit, long-nosed bats also fed on them and dispersed their seeds.

The work was grueling, but there were distinct pleasures. Volunteers were never sure what to expect next, but they enjoyed swapping tales of encounters with critters, both large and small, from desert bighorn sheep and mountain lions to pack rats, rattlesnakes, and scorpions.

An especially memorable experience for me occurred one morning as volunteers Tom and Laura Finn were helping me search for bat roosts. We spotted a split-rail corral with what appeared to be an abandoned well inside. Climbing over to check the well for bats, we quickly ran back at the sound of a snorting bull.

As Tom and Laura wisely scampered to safety, I looked back to see a really scrawny little fellow and laughed, saying, "Look what we're running from. I'll bet he's more afraid of us than we are of him." And showing a bit of foolish bravado, I turned around and rushed the bull to prove my point. With lightning speed, he charged, and I've never run harder. I vaulted the fence at full speed with a pair of sharp horns about two feet off my rear end. As the three of us stood in shocked silence at the near calamity, Laura couldn't resist grinning and asking, "So who was the chicken?" For once I had no retort.

One thing was clear. This kind of research was very different from what Mike Ryan and I had done on frog-eating bats in Panama. Instead of running experiments in which significant results could sometimes be obtained in an hour, pollination research required weeks of careful observation, combined with patient waiting to count the presence or absence of resulting fruits and seeds.

After two seasons of fieldwork, we finally had answers ready to publish.

Radio-tracking showed that remnant bats, heavily persecuted at no-longer-occupied mainland roosts, had been forced to live in caves about 19 miles offshore on Tiburon Island in the Sea of

Cortez. During the course of a night's feeding, they often traveled foraging circuits of more than 60 miles, much farther than has been reported for insect or bird pollinators.

To answer our question, "How important are bats relative to other pollinators?" our day versus night experiments, combined with direct observations, indicated that cardon and organ pipe were highly dependent on bats. These cacti opened their flowers at dusk and closed them the next morning, whereas saguaro flowers didn't open until approximately midnight and remained open until the following afternoon. Nectar production peaked in cardon and organ pipe at 10:00 PM in contrast to 2:00 and 8:00 AM in saguaro. The latter species was hedging its bets by also appealing to daytime pollinators.

Sphinx moths visited cardon and organ pipe flowers soon after they opened, but we saw no evidence of pollination. These large moths used extremely long proboscises to probe for nectar, often without contacting floral reproductive organs. In many hours of nighttime observation, lesser long-nosed bats were the only effective pollinators we saw at any of the three cactus species.

Fruit production in cardon and organ pipe was falling far short of its potential, almost certainly because of a shortage of bats.

In sharp contrast, although bats also seemed to be the most effective pollinators of saguaro, this cactus was suffering only a small shortfall. By focusing half of its energy on daytime pollinators, it was able to compensate for the shortage of bats. Nine bird species and countless honeybees visited its flowers, but the lion's share of daytime pollination was accomplished by white-winged doves. Hummingbirds were the most frequent bird visitors, but because they mostly hovered with minimal contact, they contributed little pollination. Honeybees were the least effective visitors. They confined most activities to single plants and may have had a negative impact by repelling birds.

On a per-visit basis, lesser long-nosed bats and white-winged doves were by far the most effective cactus pollinators. Given the extremely long foraging ranges of long-nosed bats, however, they likely provided the best long-distance outcrossing and seed dispersal services. This may explain why all three cactus species first opened their flowers at night.

So how did we determine that cardon and organ pipe cacti were suffering from a significant shortage of pollinators, while the shortage for saguaro was only slight? We compared rates of fruit set in hand-pollinated flowers versus control flowers, which were left open to receive pollinators throughout their entire cycles of availability. In cardon, fruit set in open-pollinated controls was 30 percent versus 73 percent in hand-pollinated female flowers. The same comparisons were 27 versus 95 percent in organ pipe and 68 versus 80 percent in saguaro.

Flowering and nectar production schedules suggested that these plants were staggering peak flowering seasons and nectar production times to minimize competition for bats, as their most valued but scarce pollinators. The cardon blooms first in early spring, followed by saguaro, with organ pipe last, placing blooming peaks of the two most bat-reliant species the farthest apart.

When it comes to attracting bats, the cardon appears to be the undisputed king among bat-pollinated cactus species. It produces many more flowers per night and per season than either saguaro or organ pipe, and its nectar is richer in sugar. We seldom saw bats visiting saguaro when cardon flowers were available.

Who is winning the competition for bats? Our study was but a snapshot in time. Winners today may be losers tomorrow. But one thing is certain: a decline in fruit production of giant cacti, which are basic providers of food and habitat for an entire ecosystem, cannot be good.

We can only wildly guess at how many thousands, perhaps hundreds of thousands, of seeds a cactus plant must produce in order to replace itself in such a harsh environment, or what the consequences of failure might be.

Fleming, his students, and collaborators have gone on to document vital bat/cactus relationships from Mexico to the Caribbean and south to Peru. They also provided the first evidence that lesser long-nosed bats migrate between southern Mexico and the southwestern United States, following a "nectar trail" of sequentially blooming and fruiting cacti traveling north and numerous agave species as they return south. A popular version of our Kino Bay findings appeared in the June 1991 issue of *National Geographic*.

CHAPTER 9

FREE-TAILED BAT CAVES AND CROP PESTS

ENTERING THE YAWNING MOUTH of Bracken Cave, 20 miles north of San Antonio, Texas, I was immediately struck by the uniqueness of the experience. It was like descending from a spacecraft onto an alien planet. I'd have choked had I not been wearing an ammonia respirator. Here in the world's largest bat colony, countless millions of dermestid beetles and their larvae feed on the droppings of 10 to 20 million Brazilian free-tailed bats (*Tadarida brasiliensis*), producing ammonia concentrations I later measured at 250 parts per million – approximately half the amount that is immediately lethal to a human. The bats survive by lowering their metabolism and allowing extraordinarily large amounts of carbon dioxide to accumulate in their lungs, buffering them from harm.

The beetles and their larvae turn the bats' droppings into a dry, constantly shifting powder in which human tracks quickly disappear, leaving the appearance of a pristine, lunarlike landscape. I felt like a first-time explorer as I descended into the bowels of the cave.

I planned to spend a month of vacation time documenting the life story of these bats as I visited three of the species' most important nursery roosts, and this was my first reconnaissance trip to assess photo possibilities. I had come at this time so I could photograph free-tailed bats giving birth to their young. Each mother bears just one baby per year, but with so many moms, the walls were covered in a solid carpet of pink, hairless pups, many with their umbilical cords and placentas still attached. At birth, free-tailed bat babies are enormous relative to those of humans. They typically weigh a quarter as much as their mothers. That's the equivalent of a 120-pound human mother bearing a 30-pound infant. Mothers also provide close to a quarter of their body weight in milk each day.

Most bats flee at the first sign of disturbance in their roosts. But because free-tailed bats are normally protected from humans by the barely survivable conditions in their caves, they are unusually fearless.

Solid mats of pink pups, mixed with mothers giving birth and nursing, covered cave walls for as far as I could see. This was a bat photographer's dream. Nevertheless, I immediately became aware that the conditions were pure hell. Humidity was high, and the body heat from roosting bats made it feel like I was approaching a furnace. Each adult bat was radiating approximately 102-degree body heat. And 100 to over 200 tons of bats can produce a lot of heat.

My eyes burned, and the profuse sweat streaming down my face blurred my vision. Then I felt like I was being stung by bees. I was being bitten by flying dermestid beetles, which inject a proteolase enzyme with their bites. The enzyme dissolves their victim's flesh, the results resembling small blood blisters. They were attacking my exposed extremities, attempting to make a meal of me.

Breathing through a respirator was difficult at best, but as I climbed down the steep entry slope, even that became increasingly challenging. Carbon dioxide from the breath of so many bats becomes trapped in lower areas, because it is heavier than air. It's not toxic, but in high concentrations it can displace oxygen in the lungs, causing a human to faint. Without prompt rescue, suffocation and death would be inevitable. The trick is to recognize dangerously heavy breathing in time to escape. Long before I wanted to end my exploration, I was forced to retreat, hoping to find safer locations for my work in Ney or Frio Caves, both located less than 100 miles to the west.

Because of the high humidity and lack of air movement, my body had no adequate means for evaporative cooling. So, when I emerged into 90-degree heat and a slight breeze outside, it felt like I'd walked into an air-conditioned room.

As I climbed back out of the sinkhole entrance, I met three men sitting in the shade of an oak tree. They were speculating on how much fun it would be to sometime throw a stick of dynamite into the cave to see how many bats would come out all at once. Of course the answer was none. They'd all be dead. These men weren't maliciously inclined. They were simply ignorant and were quite apologetic when I explained the consequences of such an act.

The appalling spectacle of how easily the world's largest remaining bat colony could be destroyed by simple ignorance provided a strong reminder of just how important public education could be. I had no idea yet that the organization I had just founded would one day own and protect this key cave and use it to educate millions of people worldwide to understand the importance of conserving bats.

The Brazilian free-tailed bat, formerly known as the Mexican free-tailed bat, is one of the world's most fascinating animals. It forms the largest colonies of any mammal, lives in gaseous environments that would kill most others, flies up to 10,000 feet aboveground, uses tailwinds for high-speed travel, adopts orphans, and forms clusters that can include up to 500 pups in a single square foot.

These bats live predominantly in Texas and northern Mexico, where huge colonies can include 10 million or more individuals. I had read reports of evening emergences that lasted for hours and could be seen for miles, and it seemed apparent that they likely were providing essential ecological services by keeping insects in check. Nevertheless, most biologists who had studied them believed that these bats were in alarming decline. I was both concerned and curious to personally experience them.

Despite the extremely inhospitable conditions I had encountered in Bracken Cave, I could hardly wait to begin taking pictures. That would begin two evenings later under more pleasant conditions. I sat on a rocky ledge in the shade of a mesquite tree, flanked by prickly pear cactus as I waited for an estimated 13 million free-tailed bats to emerge from Frio Cave. Though the temperature remained in the 90s, the dry climate, combined with an evening breeze, kept me comfortable as I was entertained by the return of hundreds of cave swallows that nested in the cave's dimly lit entry room. High on a hill, I had a fine view of the rolling hills beyond.

My first hint of a pending bat emergence came from the arrival of several red-tailed hawks and a peregrine falcon high overhead. The bats' emergence times vary from night to night, often influenced by feeding success of the previous night, but hawk arrival times are as good an indicator as any of when the bats have most recently emerged. Like other predators, the hawks find huge concentrations of bats to be tempting dinner targets.

Fully an hour before sunset, the bat flight abruptly began, almost as if someone had suddenly turned on a giant faucet. Great columns quickly stretched almost to the eastern horizon, undulating like endless, black ribbons. Then they broke into large flocks, some disappearing thousands of feet aboveground, still climbing. I wondered, who decided when to leave, and who led the individual flocks? How did they know where to go? More than 30 years later, bat biologists are still pondering such questions.

The earlier the bats fly, the more careful they must be to initially form dense columns as a hawk-avoidance strategy. When red-tailed hawks dive into the ribbons of bats, they frequently fail to catch even one, despite the fact that several bats at a time might be bouncing off their bodies as they grab wildly with their talons. I later would observe that bats often hear the wing sounds of diving hawks. In response, bat columns shuttle rapidly from side to side, preventing approaching hawks from zeroing in on individuals.

As I watched, the peregrine falcons seemed to be a bit more sophisticated at bat hunting. They would circle high overhead, swooping down at high speed on individuals that strayed alone. Once targeted, such bats had almost no chance of escape.

Pulling myself away from the awesome spectacle, I grabbed my camera and ran to a location just below the emerging bat columns. Of course they moved aside to avoid me, but when I waited motionlessly, they soon resumed their former pattern.

The resulting photos of seemingly endless bats, with a bright moon behind, were all I could have asked for. My only regret was not having any flashes ready so I could also photograph the little ring-tailed cat that I later found perched on a ledge catching bats in the cave entrance. Also known as cacomistles, these beautiful little predators are related to raccoons, from which they differ in having much smaller bodies with strikingly bushier, longer tails banded in sharply contrasting black and white rings.

No, as much as I like bats, I don't begrudge their natural predators a chance for dinner. Except when human interference tips the balance, predation is a healthy part of nature, helping ensure population health by removing the least fit individuals.

The following morning I returned on another reconnaissance mission. I needed to determine which of the three caves was best suited for the various kinds of photography I had in mind. Bracken turned out to be best for showing huge numbers of bats gathered together in single clusters. Frio was better for spectacular emergence shots, and I chose Ney for showing cave walls blanketed in baby bats close-up. I would soon learn that all three caves were miserable and potentially dangerous.

As I entered Frio Cave, in addition to my ammonia respirator, I had worn rubber knee boots, a long-sleeved shirt, and a hardhat, all saturated in bug repellent.

Assuming myself to be well prepared, I set out to fully explore the cave. The heat, as at Bracken, was nearly overwhelming, so I welcomed an opportunity to descend into a lower-level room where I found several thousand cave myotis (*Myotis velifer*).

Soon I was gasping for air as though I had just completed a fast run. Fortunately, two weeks earlier, I had reviewed a book manuscript by bat research colleague Denny Constantine, in which he had summarized risks from carbon dioxide exposure. He had warned that it was undetectable aside from noticing extra-heavy breathing. He had also cautioned that, within as little as one and a half minutes of the onset of heavy breathing, one could faint if unable to escape.

By the time awareness dawned, I was already feeling faint. Had I not read Denny's timely warning, I likely would have sat down to rest, which would have been the worst possible response. Since carbon dioxide is heavier than air, I had to climb rapidly to higher ground. I was already so dizzy that I barely made it.

My experience in Frio should have served as an early warning, but I was still ill prepared for the extreme hazards to be encountered while spending countless hours in such environments.

Ney Cave, home to some 10 million free-tails, was my next destination. Each day I'd spend at least four or five hours underground, sometimes as much as eight, so absorbed in my photography that I often forgot to eat or drink (which, in any case, was not easy to do with a respirator).

My first goal in Ney was to show mothers giving birth and nursing newborn pups. Day after day, I'd work the same areas of wall, allowing the mothers to acclimate to my presence, until I was virtually ignored. At birth, babies were enfolded in their mother's wing and tail membranes, providing only rare photographic glimpses. However, my quest did enable many fascinating observations of never-before-seen behavior. At the time, most biologists still believed that mother free-tailed bats simply "herd-nursed" whichever babies they encountered. Finding individual pups among so many was considered impossible.

Despite the confusion caused by thousands of simultaneously calling bats, and the close crowding, I often saw evidence that mothers and pups did recognize each other. As a flying mother approached, a single pup, the only one she would eventually accept, would rear up and vocalize in response to her voice. Later, colleagues Gary McCracken and Mary Gustin would conclusively document that mothers could identify their own pups by either voice or odor.

It would seem that, surrounded by millions of bats, there would be ample photographic opportunities, yet that was not the case. The constant stream of mothers flying between me and the clustered pups on the wall made picture taking difficult at best, not to mention having to constantly clean my glasses and camera lens and swat biting beetles.

To show the immensity of the bat colony in Bracken Cave, I had to wait for just the right time, combined with a slightly overcast sky. I needed the spacious entry room to be faintly lit from outside. By placing my camera on a tripod and using a time exposure, I was able to take a picture showing only the huge clusters of roosting bats. It was quite tricky getting just the right exposure length, especially in the days of film, when one could not quickly check results. The exposure had to be long enough so that flying bats would not be recorded, but short enough not to unnecessarily blur roosting bats. The resulting photo eventually appeared in my August 1995 *National Geographic* article, "Saving America's Beleaguered Bats." Had I used flashes for illumination, the tens of thousands of bats flying between the walls and me would have ruined the picture.

The next morning I returned two hours before dawn to witness the bats' return. I had read that, in its own way, this was as spectacular as the evening exodus. As I approached with a dimmed headlamp, I couldn't miss hearing staccato bursts of swishing sounds as small groups dropped into the cave entrance and abruptly slowed down.

Arrivals gradually built into a continuous stream, as hundreds of bats at a time plummeted into the cave's gaping entrance, wings half-folded, pumping to accelerate their speed, like tiny falcons. Then, just inside, they'd abruptly spread their wings to apply the brakes. What was amazing was that, with all the sudden slowdowns, from speeds I estimated to approach 80 miles per hour, no one seemed to be getting rear-ended.

Decisions on where to roost must have been virtually instantaneous in order to prevent a horrendous traffic jam. Roosting locations couldn't have all been equal. Who got the choice spots, and how were choices made? Were there leaders, and a system? Did individuals occupy the exact same places day after day?

As the sky brightened, I could see a giant but well-organized traffic jam high overhead. The sheer numbers would have given modern airport controllers a nightmare. In a 1962 monograph, Richard Davis and associates had reported that such problems at Bracken were solved by

> establishment of a landing pattern to which all groups adhered. About ten large groups formed a loose, wide, wheeling circle at an altitude estimated as exceeding 6,000 feet. As one group peeled off to plummet down, another moved in from outside the circle to take its place. Groups arriving late made long lateral flights to await their turns in the circle, but they seemed to come back in perfect time to fill an opening. This orderly array was operating so rapidly that a continuous stream of bats was entering the cave below. Each group in this flight not only maintained its integrity but also oriented in recognition of the other groups.

I could only imagine what it must be like a month later when young pups began learning to fly. On their very first flight, each must not only stay airborne long enough to reach the far wall, but must execute a perfectly timed barrel roll to get feet out in front in time to avoid a fatal collision. And all of that must be accomplished relying entirely on a previously untested ultrasonic navigation system while avoiding multiple collisions per second with other beginners.

Amazingly, most survive. On later nighttime visits I'd learn that, if I stood still in the middle of Bracken Cave, I could quickly become a practice landing pad for young fliers.

Back at Ney Cave, I was surprised to find turkey vultures entering to catch and eat baby bats. Wanting to record this novel, previously unknown activity, I watched from the interior darkness for any predictable behavior. Soon it became obvious that fallen babies, in their attempts to climb back up the wall, were sometimes climbing to the tops of large rocks instead. Such babies were doomed to die, eaten within minutes by dermestid beetles or their larvae, regardless of whether or not vultures found them. The vultures had learned to search rock tops in the dim light, and the first rocks they searched were those best located relative to fallen baby bats climbing them.

Seeing this, I drilled tiny holes in the limestone of a low ceiling and a nearby wall, then screwed ball and socket joints into them, mounted several flashes for "studio-type" lighting, and positioned a blind nearby where I could hide to take pictures. This required long, exceedingly miserable waits. I would enter the blind early in the morning, then wait for several hours for vultures to arrive, all the while attempting to hold still while dermestids and their larvae constantly climbed into my clothing, attempting to have me for breakfast.

Over a period of close to two weeks in the caves, I gradually developed more and more serious headaches, but simply took Excedrin, believing my headaches to be no more than a minor inconvenience. I gloated at how tough I was – able to withstand heat prostration and other inconveniences that would have stopped almost anyone else.

I finally did get nicely lit photos of a vulture catching a baby bat. When I returned the next morning, though, about halfway down the steep entry into the cave, I realized I was so weak that, if I continued, I might never be able to climb back out. Reluctantly, I turned around and found myself barely able to stagger back to my car. At a nearby motel, I had to enlist help to be driven into San Antonio for an emergency appointment to see a doctor.

Thinking I just had a case of the flu, I wobbled into Dr. Greg Jackson's office. He quickly informed me that I had a 103-degree fever and the worst lung X-rays he'd ever seen. He sent me straight to San Antonio's Metropolitan General Hospital, where I was tested and found to have a remaining lung capacity of just 35 percent.

I immediately attracted the attention of half a dozen doctors, most of whom assumed I must have some terrible bat disease. After all, shouldn't anyone who so closely associated with millions of bats catch a bad bug? They were quite surprised to find no evidence that I'd ever even been exposed to anything attributable to bats.

It was soon discovered that I had ammonia inhalation poisoning. My respirator had been leaking each day, gradually allowing the ammonia fumes to destroy more and more of my lungs. By taking painkillers instead of paying attention to symptoms, I'd risked my very life, a mistake I have not repeated. I also learned the importance of testing respirators.

I spent the next ten days in the hospital. By the tenth day, I was exceedingly impatient with wasting time in bed, so when Dr. Jackson came to see me, I insisted I was ready for release. He strongly disagreed, pointing out that I felt okay only because I was resting in bed. He finally said, "If you were to try doing some pushups right now, you'd see how weak you really are."

Grabbing this unexpected opportunity, I jumped out of bed and did ten one-arm pushups. The doctor couldn't help but laugh, and he released me. That evening, I treated him to a bat emergence spectacle at Frio Cave.

I apparently wasn't very clever at hiding my intentions of returning to free-tailed bat photography. So at the end of our evening together, Dr. Jackson gave me a stern warning. "Stay out of those caves! Keep in mind that even the slightest additional exposure to ammonia could result in a severe, potentially fatal allergic reaction."

I did listen sufficiently that I spent much of the following day shopping for and testing better ammonia respirators. But, inevitably, I just couldn't resist completing my mission. Two evenings later, I joined Gary McCracken and Mary Gustin in the James River Bat Cave to document their classic research showing that mother free-tailed bats and their pups do indeed recognize each other. This time I relied on a well-tested, full-face respirator.

Gary and Mary captured mothers with nursing pups by hand. They then took small blood samples for tests of genetic relatedness, and painted the tops of the bats' heads with dabs of reflective paint. They returned the bats to the cave wall where they'd captured them, and a low-light video camera with infrared illumination recorded subsequent behavior as moms left their pups while they fed, then returned to nurse.

Results from repeated mother and pup video comparisons in the cave, hundreds of call recordings, and lab tests made it clear. Each pup identified its mother's approaching voice from among thousands of others, reared up from the cluster, and called back. She also recognized her pup's voice and odor. Final recognition was confirmed when a mother sniffed and accepted her similarly painted pup. Other pups often attempted to steal a sip but were beaten back by swift cuffs from the mother's wing.

In those days, we could only guess at where mothers went to feed or what insects they ate. The first evidence came from analysis of carbon isotopes from bat feces. Native vegetation and agricultural crops have different ratios. These are incorporated into the tissues of insects that feed on them and remain measurable in the feces of bats after they eat those insects. Research teams led by David Des Marais and Hiroshi Mizutani quantified differences to determine that crop pests made up half to two-thirds of free-tailed bats' diets in New Mexico and Arizona.

In June 1995 I received a call from Jim Ward, the science and operations officer at the New Braunfels, Texas, Doppler radar facility of the National Weather Service. That call was a giant step forward for bat conservation.

Jim excitedly explained about the new radar facility, located just 19 miles from Bracken Cave, and reported, "We're having more fun watching bats than weather. The bats appear to be going directly to intercept concentrations of what we presume to be crop pests. We make nightly recordings and would be happy to show them to you."

We met the next morning. I had been pointing out for years that these bats were likely having a major impact in keeping insect populations in check, but I hadn't even dreamed that we might someday have proof that they were preferentially feeding on crop pests.

Jim introduced me to his data acquisitions manager, Bill Runyon, and they jointly offered to collaborate. I was thrilled and immediately called free-tailed bat researcher Gary McCracken. I said, "You're not going to believe the neat research opportunity I've just discovered!" When I described the radar images and Jim and Bill's offer of help, he joined in my enthusiasm, promptly flying down to meet them.

Though the radar images were quite convincing, we needed proof that the bats were actually eating crop pests, not just happening to go to the same locations. For that, Gary contacted John Westbrook, a meteorologist for the Pest Management Research Unit of the U.S. Department of Agriculture in College Station,

Texas. John was an expert on the vast migrations of moth pests that enter Texas from Mexico each spring, and he was delighted to collaborate.

Of greatest importance to Texas farmers, John hypothesized that the bats might be doing more than just eating pests that had already fed on crops. On some nights, the bats flew thousands of feet aboveground, suggesting they might be intercepting invading moths before they could lay eggs and cause damage.

John was able to obtain a mobile radar unit and weather balloons, and Gary purchased a bat detector that he tuned to the ultrasonic frequencies used by echolocating free-tailed bats. Together they connected the detector to a radio transmitter and packaged them in a waterproof container attached to a weather balloon that could be inflated with helium.

By calculating wind speed and direction, they were able to loft the rig on a trajectory that would intercept arriving moths thousands of feet aboveground. By listening to the bats' echolocation signals, they would know whether bats were feeding or just traveling. During normal travel, bats produce about 10 calls per second. However, when in hot pursuit of a flying insect, they speed the repetition rate to approximately 200 per second. The recorded sounds go from a steady beep, beep, beep to what bat biologists call a "feeding buzz." If bats were actually feeding up there, the difference should be obvious.

Accompanied by a mobile radar technician who monitored both the moths' migration and the balloon's progress, Gary and John waited in great anticipation. It was June 1996, and after extensive planning they were ready to gamble close to a thousand dollars' worth of equipment on this test with minimal probability of ever seeing their instruments again. Would it work as planned, and would they actually record feeding buzzes?

As they watched the radar screen, the technician finally announced that their balloon was entering the area of descending moths more than 2,000 feet aboveground. And as if by magic, at that very moment they began hearing feeding buzzes – many of them! The pain of losing their expensive equipment had suddenly become bearable.

The next step was to document exactly which insects the bats were eating. Up until that time, nearly all knowledge of bat feeding preferences was based on examination of tiny insect fragments obtained from fecal samples, and these rarely could be identified to species. Fortunately, new tests using DNA markers were just becoming available, enabling insect species to be identified for the first time. The bats were feeding heavily on corn earworm, tobacco budworm, and fall armyworm moths, the most costly pests of America. By eating egg-laden moths as they entered Texas in spring, our free-tailed bats were having extra impact, likely reducing crop damage all the way to Canada, as generation after generation of moths hopscotched north.

Agriculture extension agents pointed out that the bats didn't have to catch all the moths to make a big difference for farmers. Just reducing numbers below the threshold at which pesticide spraying would be necessary – even once or twice per season – could result in big savings. Pesticides cost Texas farmers more than $70 an acre to apply, not to mention poisoning our food and water.

By late 1996 we knew enough to make a remarkable case for the agricultural values of free-tailed bats. That winter, on a trip to Washington, D.C., I made a point of dropping by to visit my longtime friend Neva Folk at the National Geographic Society headquarters. She had recently become executive assistant to *National Geographic* editor-in-chief Bill Allen, and they had adjoining offices. Knowing he would likely recognize my voice and come out to greet me, I carried a tray of free-tailed bat slides, hoping to pique his curiosity about the new project.

As anticipated, he greeted me and asked what I had been up to, and I told him about the joint presentation that Gary Mc-Cracken and I had just made at the U.S. State Department on behalf of protecting key overwintering caves for free-tailed bats in Mexico. I noted that we had shown some remarkable radar images, the first of their kind to document vast insect pest migrations being intercepted by millions of Texas bats thousands of feet aboveground. Noticing the carousel tray of slides tucked under my arm, he asked, "Are those the images?"

Of course I said "Yes," and he asked to see them. Twenty minutes later, he'd also asked to cover the story in *National Geographic*. This time the article would be written by Gary and John, illustrated with some of my photographs.

Wanting to observe the moth migration and the bats feeding on them firsthand, Gary and John hoped to employ a nighttime ride in a hot-air balloon. This seemed like a perfect opportunity to photographically document their work. Not wanting to miss out on a big adventure, I hired balloonist Dave Schmuck to take us up. We rose at two in the morning and headed for an open field just a few miles from Frio Cave. Dave, Gary, John, and I were airborne by three and ascending rapidly. We couldn't launch earlier because, as Dave explained, "Nighttime landings are extremely hazardous, so we don't want to run out of fuel before dawn." We didn't want to risk striking power lines or fences.

By 4:30 AM we were drifting high above the Texas countryside. Gary, John, and I were totally enthralled with our first nighttime balloon ride. All was perfectly calm except for occasional blasts from the balloon's butane burners as we approached 5,000 feet. There was no detectable breeze because we were drifting at the same speed. Soon all sign of civilization was left behind as we drifted west, still scanning for the great flocks of moths we had anticipated, but actually seeing only scattered individuals.

In hindsight, we realized we shouldn't have been surprised. Though billions of moths showed up on Doppler radar as a huge cloud, simple math translated to just one moth per thousands of cubic feet of air space. The bats were up there feeding, all right, and they occasionally approached us to inspect this giant light bulb in the sky.

Our pilot was so fascinated by all the sophisticated equipment, from night-vision scope to ultrasonic listening and recording devices, that he apparently didn't notice when I clipped in with my safety line. It was attached to an inconspicuous seat sling I'd worn since before boarding. When suddenly he looked up and found me dangling outside the gondola, he seemed genuinely flabbergasted, despite our conversation of several days earlier in which I'd thought he understood my plan to exit the gondola to obtain better photos of the scientists at work.

"What are you doing? Get back in here now!"

My response, "But we agreed I could do this. I hang from ropes in caves all the time. I'm in no danger."

"But we're thousands of feet up, and I'm responsible." Eventually, it became clear that I was experienced at dangling from ropes, and Dave reluctantly noted that I'd been warned, and he resigned himself to my being crazy.

The nighttime ride was a fabulous experience in itself: a brilliant star-studded sky, the gradual reddening of the morning horizon, occasional groups of bats curiously checking us out, and sensational climbs over mountaintops taller than we had imagined in that part of Texas. Actually, we'd drifted well west of the relatively flat, populated farmland of the Uvalde area, and now we began to realize that we were not just in rugged terrain. It was also remote and roadless.

Then Dave announced that we had only ten minutes' worth of fuel left as we began burning even more in order to climb over yet another high ridge. Finally, in the first light of dawn we spotted a small, isolated ranch house. Descending to some 70 feet overhead, Dave yelled as loudly as possible, "Is anybody home? We need permission to land." A petite white-haired woman emerged, took one look at us, and ran back inside.

Out of fuel, we had no option but to go ahead and land. Dave did a great job of setting us down gently in the middle of 82-year-old Hallie Chism's driveway. She soon reappeared, delighted to find us all alive and okay. She'd seen me hanging from the gondola, assumed we had an emergency, and had gone to call for help.

Hallie had lost her husband a year earlier but had continued to run her ranch, herding some 400 goats all on her own. She was thrilled to welcome our surprise visit and insisted on serving us coffee and cookies. She then drove like a rally car driver while taking Dave to find our "chase car." She was a delightful character, one we'll never forget.

The groundbreaking discoveries of the value of free-tailed bats weren't forgotten either. Our article, "Bat Patrol," appeared in the April 2002 issue of *National Geographic*, followed by a whole series of scientific papers published by Gary and a team of a dozen other leading scientists, most importantly Cutler Cleveland and Tom Kunz of Boston University and Rodrigo Medellin of the National Autonomous University of Mexico. Their initial studies focused solely on these bats' impact on cotton in the

Texas Winter Garden area southwest of San Antonio. But, even in this confined study, they calculated a savings in avoided pesticide use of as much as $1,725,000 in some years from a crop valued at $4.6 to $6.4 million annually. The moths that free-tails eat also attack a wide variety of other crops, from corn and tomatoes to wheat and alfalfa. Thus, when ongoing studies that include all crops are completed, the total savings will be far higher.

To better appreciate the value of free-tailed bats, consider that during moth migrations, a single bat easily can consume more than 20 moths in a night, each carrying 500 to 1,000 eggs that otherwise would be laid on crops. A density of 5,000 to 10,000 caterpillars per acre of cotton exceeds the threshold for pesticide application. Yet 20 moths can lay from 10,000 to 20,000 eggs. If even half hatched to become caterpillars, they still could force a farmer to spray an acre of crop. The potential impact of whole free-tailed bat colonies is incredible. The cost of protecting these bats is minimal compared to the potential cost of losing them.

Today, thanks to timely research, extensive education, and generous support from numerous donors and partners, Bracken Cave is protected as part of a 2,000-acre nature reserve managed by Bat Conservation International. The story of its bats has already reached national television audiences in over 100 countries, and it is my dream to one day see this site become the world center for education about bats and their importance.

It is a matter of great personal pride that the three men I originally found discussing dynamiting Bracken Cave later volunteered their help in habitat restoration; and that the Texas Department of Transportation, which formerly exterminated bats found in bridges, now proudly incorporates bat habitat into new bridges in agricultural areas. Such bridges shelter millions of Brazilian free-tails throughout much of the species' United States range, thanks largely to the personal efforts of bridge engineer and BCI member Mark Bloschock.

Worldwide, there are more than 100 species of free-tailed bats, and they live on every continent except Antarctica. Several that live in caves form the largest aggregations of any mammal, especially in North America, Africa, and Asia. Evidence of their agricultural contributions in America has triggered subsequent studies that suggest similar benefit in protecting major crops in Asia and Africa, from rice and sugar cane to macadamia nuts.

Though free-tails are now protected in a majority of their largest United States roosts, these invaluable allies still face an uncertain future. Many of Mexico's free-tails have suffered alarming decline in recent decades, and even United States colonies, whose roosts are now protected, remain vulnerable to carelessly located and operated wind energy facilities. In Asia and Africa, many formerly great colonies have lost their caves to limestone quarrying. Only as more people learn of their importance will the future of free-tailed bats be truly secure.

CHAPTER 10

AFRICAN ADVENTURES

CLANKING METAL POLES and singing lustily, "A-hunting we will go," I followed a game trail through dense, head-high reeds. It was 2 AM on my first night in Africa, and I was determined to reach the Sengwa River, where I had earlier set a 42-foot-long mist net over the water.

Just before sundown, two armed game wardens had escorted me to this location where fellow bat researcher Brock Fenton had reported catching a male Chapin's free-tailed bat (*Chaerephon chapini*), one of Africa's least-seen mammals. About a third smaller than my thumb, they belong to the free-tailed bat family (Molossidae), relatives of our Brazilian free-tailed bats in Texas.

I had invested $10,000 of personal funds in a trip to Zimbabwe, and now I was hell-bent to capture and photograph this species in which males have a showy crest that they can raise the way a peacock spreads its tail. The game wardens had waited with me until 11 PM, but we hadn't seen a single bat of any kind. The wardens had normal daytime work to do, so I couldn't ask them to stay later.

After closing the net and returning to the Sengwa Wildlife Research Area headquarters, I had noticed that the moon would soon set. I went to bed, but couldn't sleep. I was aware that bats often escape owls by avoiding moonlight, and feeling desperate after my initial failure, I got up and headed back down to the river.

I was now alone, unarmed, and more than a little fearful of running into one of the many elephants or Cape buffalo we'd seen the evening before. The game wardens had assured me that both preferred to avoid humans, but they'd also not suspected I'd be crazy enough to return by myself. I was making all the noise possible and hoping for the best as I worked my way through an area where I couldn't see ten feet in front of me through the dense vegetation.

When, half an hour later, I arrived at my net, I faced a new set of challenges. The night was surprisingly chilly, and I had to wade waist-deep into the river in order to open and check my net. Soon there was nothing to do but wait, listen, and shiver in the cold river as the moon disappeared. Nearby were the steady honking sounds I would later come to recognize as a courting male epauletted bat, and farther off the macabre laughs of hyenas. Though I heard many new sounds that first night, it was the lions that I'll never forget.

I had been warned to periodically scan for crocodile eyes, but no one had mentioned lions. Just as I had become chilled to the bone, I heard what even a novice in Africa could deduce to be approaching lions — and not just one or two, but apparently a whole pride. The only available shelter was beneath an undercut in the riverbank where the water was deepest. Not wanting to face the lions, I took refuge there as quickly and silently as possible. Soon, the lions were almost directly overhead, apparently having decided to hang around for a while.

I hardly dared move for fear of being discovered, but then the aquatic insects Brock had warned me about began to bite, each one feeling like a wasp sting.

The lions stayed for what seemed the longest hour of my life. No crocs arrived, and the sun finally rose. By that time I was nearly hypothermic and still hadn't caught a single bat. The wardens weren't amused to hear what I'd done but were good sports about helping again.

The next night, at a new location, I was thrilled to catch a male and two female Chapin's free-tailed bats by setting two nets, one above the other, on extra-tall poles, and I was especially grateful to catch them before the game wardens had to leave.

I eventually got an impressive portrait showing the male with his crest spread. To this day no one knows much about how these crests are used, but we presume they're spread during courtship, likely accompanied by singing. Through the observations of bat rehabilitator Barbara French, we do know that male Brazilian free-tailed bats sing to attract females. Perhaps Chapin's males combine crest-spreading with singing. We have much yet to learn about courtship in bats, but singing is widespread, perhaps as common as in birds.

Three weeks after leaving Zimbabwe, while photographing lesser mouse-tailed bats (*Rhinopoma hardwickii*) near Lake Baringo in Kenya, I learned of a new hazard. Locals had informed me of a nearby bat cave. This time, I was in a relatively civilized area where I assumed – incorrectly – that I'd be fine on my own.

Once at the cave, I realized that there was a much shorter route back to my vehicle and wondered why I hadn't been told about it. If I went straight down the narrow canyon, the trip back to my vehicle would be only a couple hundred yards. I didn't consider that there might have been a good reason for taking a half-mile-longer route. Though mildly nervous about cobras and mambas, I began forcing my way through about 100 yards of waist-high vegetation. I didn't consider that the plants themselves might be a problem.

Suddenly, my bare legs were afire, and I initially had no idea why. I'd never before encountered stinging nettle with a delayed reaction, but now that I was in the very middle of it, there seemed little alternative but to continue forcing my way through. Soon it took little imagination to envision hell!

It didn't take long to figure out that there were countless reasons why one should hire a dependable guide when traversing foreign territory. I contacted well-known Kenyan wildlife artist Rob Glen for his advice in finding a reliable guide.

He highly recommended Paul Kabochi and organized an opportunity for me to meet Paul at Rob's home near Nairobi. My first impression couldn't have been more wrong. Paul was wearing pants and an old, threadbare sports jacket that looked like something he'd rescued many years earlier from someone's trash pile. This, combined with high-heeled shoes, made him look utterly ridiculous. He was about five feet, six inches tall and skinny as a rail. He was Kikuyu, a member of Kenya's largest ethnic group.

Rob had explained that Paul was the best naturalist he'd ever met, and also that he had been trained by the Israeli Mossad in counter-terrorism and intelligence. He was said to have been extremely successful in infiltrating and destroying Somali guerrilla bands while in the Kenyan military. Given his training, he'd be especially able to keep me out of trouble.

But on our first meeting, I couldn't believe Paul could protect me from anything. He was so small and fragile-looking that it seemed even a soft breeze could have blown him over. I knew Rob was a practical joker, so I steeled myself not to be taken in. Reading my thoughts, Rob just grinned and said, "It would be a big mistake to ever underestimate him. He's as honest and hard-working a man as you'll ever meet, but he also can take good care of you. I've often relied on him myself."

That was hard for me to believe, since Rob regularly worked out and didn't exactly look like he'd ever need protecting. I finally agreed to give him a try. That was one of the best decisions I ever made.

Paul never ceased to amaze me. He spoke 13 languages, could find almost any animal in minutes, and had a special way with people. He also knew bats and how to find them.

Our first field location was Hunter's Lodge, an oasis vacation spot for Kenyans. Paul didn't seem capable of relaxing. He was constantly scouting new bat sites, helping me take pictures, or washing our Land Rover if he ran out of other things to do. As a result, we were able to photograph eight species of bats in just the first few days.

A large lava tube cave several kilometers away contained a colony of over 100,000 large-eared mastiff bats (*Otomops martiensseni*). Large clusters covered high ceilings, and the floor was covered in dermestid beetles and their larvae, very similar to what I had experienced in Bracken Cave in Texas.

Large-eared mastiffs belong to the free-tailed family, having tails that extend about an inch beyond the tail membrane. They are related to the western bonneted bats of North America but are only about half as large. They also are more colorful and have larger ears for their size. Their shoulders are dark chocolate to rusty red with lighter bodies bordered by white along slender wings. Like other free-tailed bats, they are speedy, long-distance fliers that often form large colonies. Their sturdy ears may serve as air foils to provide lift during high-speed flight.

That evening, while I photographed one of these bats back at my studio, Paul set mist nets nearby and caught several other species, the most interesting being the first Wahlberg's epauletted fruit bat (*Epomophorus wahlbergi*) I'd seen close up. Males were courting in nearby fig trees where they were performing wing dances – singing through inflated cheek pouches and flashing their shoulder epaulettes.

The large male that Paul brought me loved to eat ripe figs from my hand, and I soon learned that not only were his pouches useful for enhancing his songs, but, similar to hamsters or chipmunks, they could also be filled with food. Intrigued, I took a series of photos of him stuffing a large fig into his pouches. One of these photos would become one of my most popular bat pictures. Audiences the world over would consistently be amused by it.

The next morning Paul found me a small family group of banana bats (*Neoromicia nanus*) roosting in an unfurling banana leaf. These tiny insect-eaters weigh slightly less than a United States nickel and have small adhesion pads on their wrists and ankles that aid in climbing slick-surfaced leaves. This is just one of more than a half-dozen kinds of bats, scattered throughout the world's tropics, that have independently evolved special structures for clinging to the insides of unfurling leaves.

Paul was a miracle worker when it came to finding even the most difficult species. The yellow-winged bat (*Lavia frons*) was one of my top priorities for illustrating the beauty of bats. Their light bluish gray fur is long and soft, and the wings are as bright yellow as any tanager. Males even have yellow ears and noseleaves. They're simply stunning.

These bats feed like flycatcher birds. They hang from a strategically located perch, pouncing on large insects as they pass. They use their extra-large ears to detect wing-beat sounds and footsteps. Yellow-winged bats live in monogamous pairs, and males appear to check out feeding locations early in the evening, sharing information about the best places with their mates.

Despite my intense desire to photograph these extraordinarily colorful bats, I had been warned by bat research colleague Terry Vaughan that they were nearly impossible to capture, and Rob Glen agreed.

When I told Paul how much I'd like to catch a yellow-winged bat, he said, "No problem," his usual response to being told something was difficult or impossible. At our next destination, down in the famous Rift Valley, he explained that we'd have to set a mist net in front of a bright floodlight in the yard at the camp where we were staying. Had I not already developed a great deal of respect for his wisdom, I'd have told him he was insane. Any bat researcher knew better than to set a bat net in a brightly lit location where most bats would easily see and avoid it. Nevertheless, with a great deal of skepticism, I helped him set a net in just such a spot. I didn't even wait to watch.

Needless to say, I was astonished when, less than an hour later, he knocked on my door, proudly bearing not one but two yellow-winged bats, his only scornful comment, "Easy!" Soon, finding him invaluable, I doubled his wages.

Having seen what appeared to be some ideal research opportunities during my first visit in 1982, two years later I invited Mike Ryan to join me in a project to see if any of Africa's carnivorous bats were relying on frog calls to hunt prey. Again funded by the National Geographic Committee for Research and Exploration, our first priority was to recruit Paul Kabochi as our assistant. Paul was available and delighted to help us – lucky for us, because he probably saved our lives.

We were near Msambweni, about 50 miles south of Mombasa, recording frog calls. It was nearly 11 PM, and Mike, Paul, and I had just returned to our Land Rover from working along the Ramisi River. We were preparing to drive away when we suddenly found ourselves surrounded by a dozen African tribesmen. Armed with spears and bows and arrows, they clearly weren't the kind one would wish to meet. They were speaking a local dialect that they wrongly assumed none of us could understand, since Paul cleverly spoke to them only in Swahili.

As Paul listened, it became apparent to him that these men were bandits. They were openly discussing how to rob and kill us. Luckily, Paul not only understood, but was well prepared to deal with such an emergency, thanks to his counterinsurgency training by the Mossad.

Now, with no time to explain, he simply said "Keep the engine running" and jumped out, leaving the door open. Astonished by his bold move and sudden ability to speak their language, the bandits immediately gathered around Paul on his side of the vehicle. As their leader attempted to provoke an argument, those behind Paul gradually edged away, signaling to Paul that he was about to be attacked.

In the nick of time, he further surprised them by feigning an attack on their leader. Then, just as they were all caught off-guard and were off-balance leaning back, he dived back into the vehicle, grabbed our powerful blaster light, blinded them with it, and yelled "Go!" Mike needed no encouragement. No longer surrounded, he floored it, and we escaped as spears thudded into the rear end of our vehicle.

The next morning, assuming we'd never dare venture into the local countryside at night again, Mike and I were discussing our options when Paul interrupted with one of his standard "No problem!" statements. Shocked, Mike and I immediately responded in unison, "We certainly do have a problem! We wouldn't dare go near that place again."

Paul just laughed and said, "You don't understand Africa. When you're ready, I'll take you to meet the Msambweni sub-chief. His name is Mr. Bakari Mahaba." At Paul's insistence, we went back to meet the sub-chief. Paul showed him great respect, and he listened carefully as Paul explained our problem of the previous night. He spoke only a few words to Paul in response, and then asked us to take him back to the site of the problem.

Of course we complied, and he introduced us to the elders in the nearby village, explaining that if anyone bothered us again, he would have to send in the police to investigate. As we drove the sub-chief home, he gave us his personal guarantee that we could return anytime in complete safety.

After bidding him a polite farewell, Paul explained how he could be so certain. The police were much feared. If called to investigate an incident, they would ask who was involved, and if the villagers didn't promptly identify as many culprits as had been reported, they would beat anyone who might know anything, including women and children, until someone confessed.

Though scientists like myself often enjoy recalling scary escapades, in reality, we never had more fun than in the next three months in Kenya. We had rented a comfortable four-bedroom bungalow on the Diani Beach, where we lived like kings while conducting our research without further incident. The shilling in Kenya had just been devalued, so our grant went nearly twice as far as anticipated.

For dinner the evening after our scary episode, we celebrated with a five-pound lobster, followed by a delicious dessert of fruit salad composed of mangoes, pineapple, papaya, and bananas with fresh-roasted cashews served on the side. A wide variety of seafood and tropical fruit was delivered to our door daily. Life was good.

Soon after our arrival, Mike and I began searching local caves for the heart-nosed bats (*Cardioderma cor*) we had come to study. They were living in small, nearby caves and were easily caught in mist nets set at entrances. We were alarmed, however, to find that nearly half of the area's caves had been permanently destroyed. Several of the larger ones had been converted for use as restaurants or other tourist attractions.

One cave, reported to have formerly sheltered a large bat colony, had recently been closed by the owner after he was misinformed that if he didn't kill the bats they would soon invade his new home. The species involved never lived in buildings. Other landowners had feared their livestock would fall in or that their children would get hurt. None saw any advantage to saving bats, despite the fact that most of the bats were insectivorous, probably benefiting local crops.

Though I wasn't initially happy to find witch doctors disturbing remaining bat caves, I came to respect the fact that some of those caves likely would have been destroyed had it not been for witchcraft. We could always tell which ones had been used for such ceremonies by the circularly arranged bottles, white chicken feathers, and fresh blood.

The most important remaining bat cave was called the Similani Cave. It was still occupied by about 100,000 Egyptian rousette fruit bats (*Rousettus aegyptiacus*), clearly the most important seed dispersers in an area where reforestation on abandoned land was sorely needed. It also was home to some 20,000 insect-eating bats of several species.

The cave was one of the most beautiful I've seen, not because of stalactites, stalagmites, or other formations, but because of its crystal-clear, multicolored tidal pools, high-domed ceilings with skylights, and long, hanging vines. Humans could enter only during low tide. The area's most influential witch doctor regularly used it. He inadvertently afforded the bats some protection by threatening to put curses on people who entered without his consent.

Unfortunately, bat researchers didn't qualify among his permitted guests. When he threatened to curse us, Paul simply explained that we were capable of some of the most powerful witchcraft he'd ever seen, and that the witch doctor should be very careful not to offend us, as we were capable of reversing curses to fall upon him and his clients. Thereafter we were studiously avoided. Years later, when Paul and I brought Dieter Plage to film there for the documentary *The Secret World of Bats,* the old man apparently still remembered us, as he quickly fled on our arrival at the cave.

Back to our research goals, I had noticed during my first visit to Kenya that heart-nosed bats were very similar to the frog-eating bats that Mike and I had studied in Panama. They were almost exactly the same size. Both had extra-large ears and relatively short, broad wings for helicopter-like flight, but they had independently evolved from two different families of bats: the Old World Megadermatidae and the New World Phyllostomidae.

Both were believed to be at least partially carnivorous, and both lived where frogs were available. If heart-nosed bats had also developed an ability to identify frogs by their mating calls, it could be exceptionally interesting to compare their independently evolved abilities.

We set up my enclosed studio in one of our cottage's spare bedrooms, tamed the bats to feed from our hands just as we'd done with frog-eating bats in Panama, and were encouraged to see that they too enjoyed eating small frogs. There was, however, one striking difference. Unlike our bats in Panama, these did not seem capable of finding frogs by their courtship calls. When we played locally recorded calls of the very species they loved to eat, they'd get excited and hunt, but they didn't find frogs placed on our speakers.

Within a few nights we were able to demonstrate that heart-nosed bats simply could not hear the low frequencies of frog calls sufficiently well to locate them, nor could they rely on vision even when close.

We designed a simple experiment to test their ability. We would tie a euthanized frog to a long string and pull it along a two-foot-long piece of glass about four inches wide. If we fastened brown paper from grocery bags on top of the glass and pulled a frog along on the string, both low- and high-frequency sounds were recorded on our sensitive microphones, and a bat would immediately come down and grab the frog. If we repeated the same experiment with the frog directly on the glass (moistened to prevent sound generation), now with the paper below (to look the same), no sound was recorded, and the bats ignored the moving frog.

Our heart-nosed bats seemed to be hunting solely based on high-frequency sounds produced by prey movements, despite the fact that this species is well known for having extra-large eyes. These bats use echolocation during their approach, but apparently not to detect prey.

To test the hypothesis that these bats were relying solely on high-frequency sounds produced by prey movements, we repeated our experiment, this time with the paper taped to the top of the glass, but only for the first six inches. When we'd pull the frog along the sound-producing paper, a bat would immediately begin its approach. But if the frog was pulled onto the wet glass, just before the bat arrived, the bat would land where it had last heard the frog moving on the paper and appear to search. None of our bats ever successfully found a frog once it was soundless on the glass, even if it was close, plainly visible, and continued to move.

We repeated these experiments at varied light levels with the same result. Our bats apparently could not rely on either echolocation or vision to find their prey. We published our results in the *Journal of Comparative Physiology*, explaining why these bats can't specialize in frog-eating, despite their ability to eat frogs.

Fieldwork in Africa was full of additional surprises. Normally we were quite cautious about crawling into hollow trees where one might easily encounter a cobra or a mamba. At one hollow baobab tree, I scanned the ground level inside for snakes but failed to look up far enough before crawling in. My attention was focused on the far wall, where I'd spotted an Egyptian slit-faced bat (*Nycteris thebaica*). I invited Mike to come have a look.

As he entered and sat, we suddenly heard a loud buzzing sound. Looking up, we spotted a large swarm of African bees less than two feet over our heads. Mike must have set a world record for the speed at which a human can make a U-turn, and I was urging him to move faster.

Mike had another surprise one evening while using our night-vision scope to observe calling frogs at a small pool in the Rift Valley. He was comfortably leaning against a sloping rock, night-vision scope and infrared light in hand, when he noticed a long shadow slowly advancing across the pond. When he looked up to spot the source, he realized that a giraffe was approaching.

He hadn't heard it above the din of calling frogs. But now he didn't know whether to stay still and risk being stepped on, or move and potentially get kicked by the surprised animal. Fortunately, it moved on before he had to do anything. About that time, we heard a lion and decided we'd seen enough.

A month later, while looking for bat caves in Tsavo West National Park, I made a foolish mistake. Paul was driving, while Mike and I enjoyed wildlife watching from the back of our open-topped Land Rover. We'd driven right by several small herds of elephants without incident, but then we encountered a group that seemed to be in a very different mood. These, though hundreds of feet away, looked aggressive. Always curious about animal behavior, I asked Paul to stop the vehicle to see what would happen. Immediately the lead cow charged.

Given what seemed like a safe distance, I told Paul to keep the engine running and in gear so I could photograph the charge. All I could think of was the great photo op. As she charged, I got so carried away with the beautiful scene – charging elephant in the foreground with a herd behind – that I waited a bit too long to yell "Go!" Paul, realizing the elephant was now only a second away, popped the clutch and stalled the engine. His driving ability wasn't as well honed as some of his other skills.

Mike and I dove to the floor just as the elephant hit the rear end of our vehicle. Eye-to-eye with an enraged elephant, we just froze and stared back. Then, as we feared the worst, the striking elephant caused our vehicle to lurch forward, starting the stalled engine. Though she continued to chase us, we gradually pulled away from her. As soon as our escape was clear, Mike jumped up, grabbed me by the shoulders, and began yelling, "Merlin, can you believe it? We're alive! We're still alive!"

In my experience, most elephants have been amiable, certainly not inclined to charge from a quarter of a mile away. This group must have been having a very bad day.

Highly recommended by Mike and me, Paul quickly became a much-sought-after guide, expert animal trainer, and researcher for field biologists and filmmakers, including the BBC and National Geographic. He continued to periodically work with me on a variety of projects through 1996 and never ceased to impress all who knew him.

Several times Paul helped me lead eco-tours for conservation donors. Not only could he find rare bats, but he was also an expert handler of poisonous snakes, from Gaboon vipers to spitting cobras, and he appeared to work near miracles with his knowledge of medicinal plants. When an 85-year-old man fell and badly scraped his arm on sharp rocks, Paul said there was no need to rush him to a doctor. The wound promptly healed after he gently rubbed a gooey mixture of mashed plant leaves on it. When he heard that a couple had been unable to have children, he insisted they both drink an obnoxious herbal tea he made from local plants. Soon after, the wife became pregnant with twins. Virtually everyone who knew Paul has a memorable story to tell.

We all especially remember him for his frequent warnings about the dangers of working around elephants. Ironically, despite all his special understanding and warnings, it was an elephant that prematurely ended his life on February 11, 2003. He was working for a film crew, tracking dwarf mongooses alone on his 61st birthday, when he was trampled and killed by a lone elephant.

▶ Painted bats blend with the orange color of wilted banana leaf tips where they roost.

◀ A dish-shaped sea bean flower petal reflects bat navigation signals, guiding the bat to a tiny trigger that fires pollen onto its rump from spring-loaded anthers.

▶ Dwarf epauletted bats, along with several other fruit-eating bats, are essential seed dispersers in West Africa, accounting for up to 95 percent of aerial seed rain in cleared areas.

◀ Gray-headed and other flying fox species are key pollinators of Australian forests, including many trees of great economic value. Yet these bats are in alarming decline, frequent victims of grossly exaggerated human fear.

▶ [This bat is deliberately turned upside down for human impact (right side up to us!).] An adult male epauletted bat has filled his cheek pouches with wild figs in Kenya. When the bat is not eating, these pouches can be inflated to enhance courtship songs.

▲ Egyptian fruit bats are often killed as pests by mango growers, but the author's Kenya study revealed that by eating only mangoes that ripen prematurely or are missed by pickers, the bats help protect crops from fungi and fruit flies.

▶ Greater mouse-eared bats of Europe hunt prey exclusively by listening to footsteps or courtship calls. This one has detected a katydid.

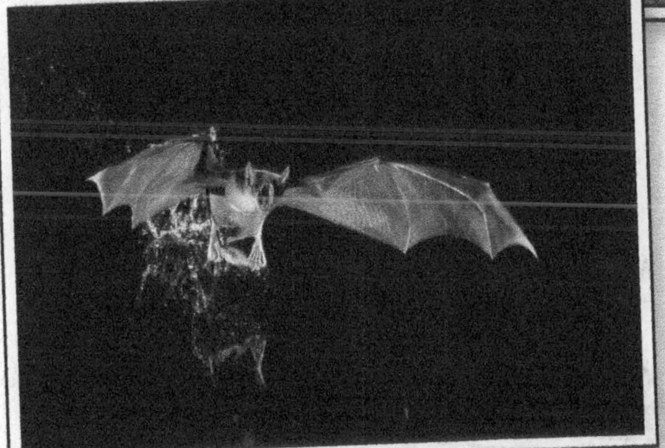

◀ Greater fishing bats of Latin America rely on huge feet and flattened toes and claws to snatch minnows from ponds.

▶ Honduran white bats use their teeth to cut midribs of large heliconia leaves to form roosting tents. Groups normally consist of a male and his harem of females and their offspring.

▶ Hardwicke's woolly bats of Borneo prefer to live in bat-adapted pitcher plants that provide shelter in exchange for feeding on their nitrogen-rich droppings.

◀ Spix's disk-winged bats emerge from their roost in an unfurling heliconia leaf (Panama). Adhesive disks on their wrists and ankles help them cling to slick leaf surfaces.

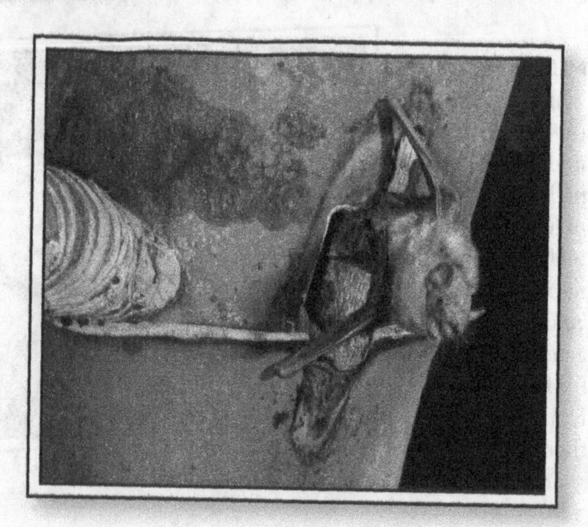

▶ Tiny bamboo bats have flattened skulls that enable them to quickly enter and exit their homes in bamboo stalks through beetle holes (China).

◀ Large flying foxes have average wingspans of approximately five feet, though extra-large individuals may reach six feet. They are severely threatened by overharvesting for human food, a problem that can threaten whole ecosystems and local economies.

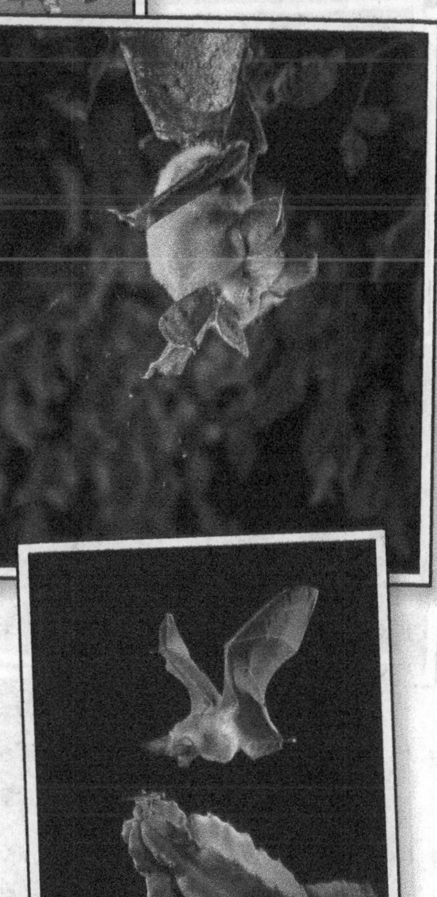

▶ Horseshoe bats of the Old World hunt with extra-high-frequency echolocation signals to avoid detection by moths, whose ears are tuned to the mid-range frequencies used by most other bat species. A few uncommon bat species can avoid detection by using extra-high or low frequencies.

▶ California leaf-nosed bats can survive in America's hottest deserts for months at a time without drinking a drop of water and can hear even the faint footsteps of crickets.

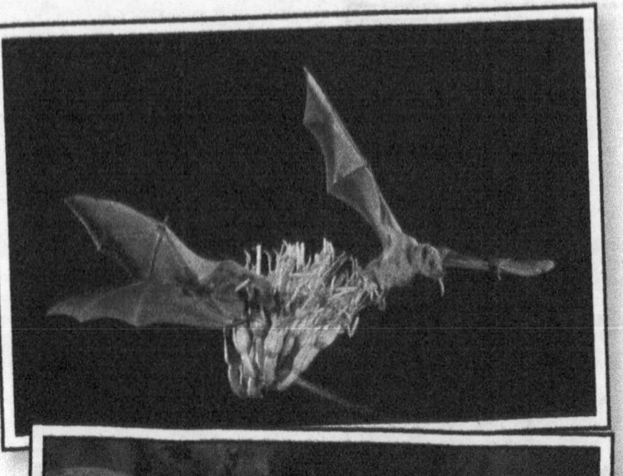

◀ The agave from which all tequila is made relies on long-nosed bats for pollination.

◀ Pallid bats live in arid areas of western North America and are immune to even the deadliest stings of the scorpions and centipedes upon which they feed.

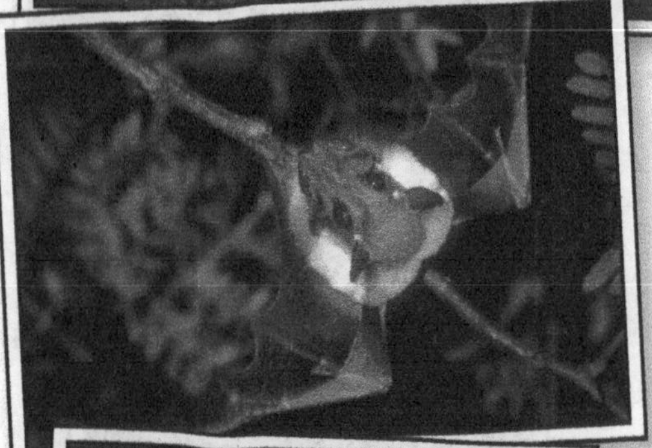

◀ Male epauletted bats inflate their cheek pouches to enhance courtship calls, flash white epaulettes of fur from their shoulder pouches, and do a "wing dance" to lure potential mates.

◀ Cuban flower bats are primary pollinators of the blue mahoe tree, prized for its timber.

▲ Bats, such as this Leach's long-tongued bat, are ideal pollinators for rare, widely scattered plant species such as this echo vine in Cuba. The dish-shaped leaf above is an echo reflector that guides approaching bats. The red nectaries that hang below reward bats with copious nectar.

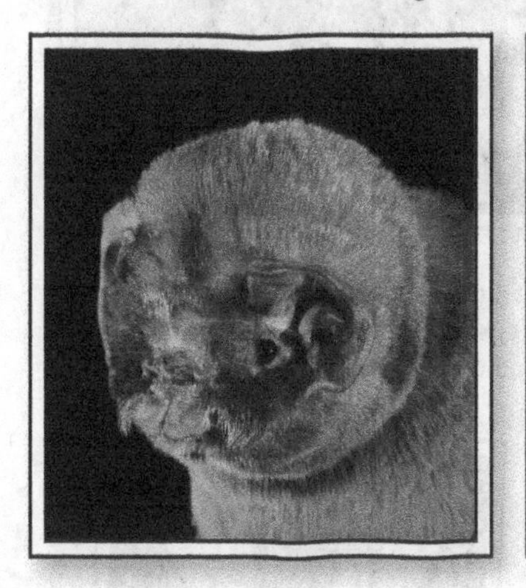

▲ Ghost-faced bats have eyes that appear to be located in their ears, perhaps accounting for their unusual name. They are insectivorous and form large colonies in caves from the southwestern United States through most of Latin America.

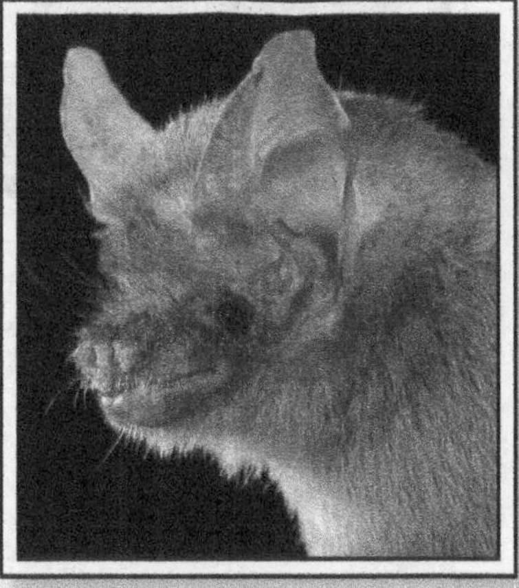

▲ The Kitti's hog-nosed bat feeds on tiny insects, weighs less than a United States penny, and is widely considered to be the world's smallest mammal. It lives in caves in Thailand and Myanmar.

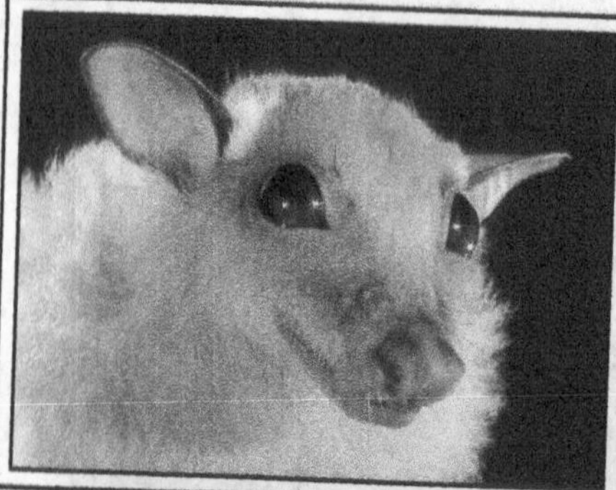

◀ Epauletted bats are essential pollinators and seed dispersers across wide expanses of Africa. They live in small colonies in tree foliage.

▲ Yellow-winged bats live in monogamous pairs. They use large ears to listen for footsteps or wingbeats of potential prey. Their bright colors blend with surrounding foliage in African savannahs.

▶ In Latin America, one short-tailed fruit bat can carry 60,000 seeds to new locations in a single night, having a huge impact on dispersal of "pioneer" plants into clearings.

CHAPTER 11

BAT-LOVING MONKS, TIGERS, AND POACHERS

PEERING THROUGH CLOUDS of vehicle exhaust so dense that Verne Read and I could barely see a hundred feet ahead, our expert driver inched through downtown Bangkok, weaving in and out of trucks, rickshaws, bicycles, and motor scooters that squeezed within mere inches on all sides of our car. We were guests of Dr. Boonsong Lekagul, Thailand's leading conservationist. He had invited us to visit, hoping that our investigation, combined with my photographs, could help stem the decline in his country's bat populations.

We were now leaving Bangkok, en route to our first destination, the majestic Rakang Cave with its 60-foot-tall cathedral-shaped entrance. Dr. Lekagul had seen an expanding limestone quarry nearby and feared for the cave's future. Two hours later, as we arrived in the entrance, we were welcomed by a dozen colorfully dressed women. Each had a small, homemade broom with no handle. Squatting low, they were sweeping a mixture of dry, powdery bat droppings and small pebbles into piles.

As mounds accumulated, they were scooped up and sifted through flat, porously woven strainer baskets into larger, rounded baskets. The women balanced the baskets on their heads and nimbly picked their barefooted way through large rocks to a flat area near the entrance. There they emptied the baskets into large bags to be carried out by men dressed only in loincloths.

Despite the dirty nature of their labor, enveloped in clouds of dust, everyone seemed happy, swapping stories and laughing as they worked.

Suddenly, my thoughts were shattered by a loud boom and what felt like an earthquake. Several rocks fell from the ceiling, but fortunately no one was hit. The guano miners explained that the quarry had advanced through the back of the hill to within less than 100 feet of their cave. We'd just experienced one of their blasts.

The cave had sheltered more than one million bats when Dr. Lekagul had last visited several years earlier, but a major decline in bat numbers had already occurred. Villagers blamed blasting disturbances and poaching by quarry workers. At current rates of limestone extraction, the cave could be entirely destroyed within months. Fifteen families had been extracting and selling bat guano for more than 100 years. Now the only livelihood they knew was at risk.

We were horrified but there seemed little we could do. Dr. Lekagul lamented that this was not a unique problem, that other caves sheltering large bat populations already had been destroyed. He noted that bat guano extraction for fertilizer provided a whole village of miners with a sustainable income that was actually much better than that of the few quarry workers who would have to abandon the site as soon as the limestone was exhausted. If only the quarry operators could leave the relatively small cave area intact, a whole village could be saved, not to mention ensuring continued pest control over surrounding rice paddies.

As we drove another 100 miles northwest to Sai Yok National Park on the border with Myanmar (Burma), I couldn't get the villagers at Rakang Cave out of my mind.

The park had been created in celebration of biologist Kitti Thonglongya's 1974 discovery of the bumblebee bat, a bat so unique that a new family, genus, and species had to be named to accommodate it: Kitti's hog-nosed bat (*Craseonycteris thonglongyai*) in the Craseonycteridae family. This bat is widely believed to be the world's smallest mammal, though several other bats and a shrew are debatably close competitors. Adults weigh a third less than a United States penny.

The park had been created specifically for this species, but ideas about bat conservation were primitive in those days. Well-marked trails led to the bats' originally best-known roosting caves, and unsupervised visitors had long since disturbed the bats into leaving.

Having failed to find bumblebee bats in the park's officially publicized roosting caves, we were led to a remote cave that people avoided because a tiger sometimes lived there. For this trip we were assigned a dozen rangers, each armed with a 16-gauge shotgun. At first, our trail wound among 100-foot-tall clumps of bamboo. Then we were plunging through dense, vine-filled undergrowth where I became quite nervous.

The rangers were short and carried their shotguns slung over their shoulders on leather straps, forcing me to stare directly into muzzles in front of me. When we paused to rest, I signaled my concern about possibly being shot. The guard next in line immediately understood. Smiling reassuringly, he pulled a three-foot-long oiled rag out of his shotgun barrel, and I began to reconsider the threat of tigers. Obviously my protector's gun could not have been fired with a rag in it.

When we reached the cave, we carefully approached from one side and scanned the soft earth for tiger tracks. Seeing none, I switched my headlamp on and slowly entered, leaving Verne and the guards behind so as not to frighten the bats. I'd gone only about 50 feet when I heard the unmistakable sound of a sizable animal. I immediately crouched against a wall, no longer confident I hadn't missed a tiger's track. I didn't dare move as I attempted to see around a turn in the passage. What was probably only seconds seemed like an hour as I waited to see if the rumored tiger was actually home. Soon after I stopped moving, it gradually came closer. Straining my eyes to spot possible tiger eyes, I was quite relieved to see my first Malayan porcupine. It had apparently felt cornered and only sought a route for escape. When I illuminated it in my headlamp, it rattled its impressive quills and then waddled on by. Unfortunately, there were still no bats.

Next, we hired a motorized canoe to take us down the Khwae Yai River (the River Kwai) in search of a larger cave guarded by a monk as a Buddhist shrine. Such shrines are common in Thailand, occasionally at great detriment to bats, but sometimes also providing extraordinary protection. In this cave, we discovered the only colony of bumblebee bats known to remain.

Such tiny bats could not be photographed in the wild without undue disturbance to the colony. Nevertheless, we needed photos in order to publicize their plight. We set a small, fine-threaded mist net to capture one of those flying by and quickly caught two.

Handling them was nerve-wracking. They could be injured or killed by the slightest mistake, and as the founder of Bat Conservation International, I felt special pressure to do no harm. With the greatest care, we carried them back to the park headquarters in a damp cloth bag to prevent overheating or desiccation.

A room had been provided for my bat photography, but I've never been more nervous. These bats were high-strung, constantly trying to escape, and each time one flew I had to catch it quickly in a butterfly net without injuring it. Thai park officials were seated next to me, watching to be sure the bats weren't harmed. And every time a bat moved, or failed to move, one of them would warn me that it might be about to die. But tiny bats with tiny eyes are among the world's most difficult animals to photograph. I wouldn't know if I had gotten suitable pictures until I could see processed film at home. (Fortunately, I no longer face such uncertainty, as for the past decade I've exclusively relied on digital photography.)

Getting close-up portraits was especially difficult, partly because I had no mealworms with which to bribe them. Expecting to be eaten themselves, the bats would squint their eyes shut or bare their teeth in self-defense. A couple of hours later, I was quite relieved to finally allow both bats to fly away free.

The next morning, we met with Thailand's director of parks, who was quite sympathetic. He promised improved protection. (Presumably, our suggestions to better protect their roosts were helpful, because when I returned many years later we found several healthy colonies.)

After two days in the national park, Dr. Lekagul had to return to his busy medical practice in Bangkok, accompanied by Verne, who had family commitments back home. For me, though, this was only the beginning of a more than month-long trip.

I rented a vehicle, and Dr. Lekagul helped me hire an assistant, Surapon Duangkhae. Surapon was an exceptionally bright and personable college student, and we quickly became friends. At the time neither of us could have anticipated the long-term impact of our time together.

Our first destination was Khao Chong Pran Cave, located in Thailand's Ratchaburi Province, just a few miles from Rakang Cave. This large cave was owned by a Buddhist monastery, where monks sold bat guano fertilizer as their primary source of income. The bats had supported the local community for as long as anyone could remember; however, the monastery's meticulously kept records showed that over the last five years, guano production had fallen by close to 50 percent, and the monks were concerned.

Dr. Lekagul had sent us there simply because he suspected it would be a good place to obtain bat photos needed for educational materials. He wasn't aware that these bats were also in trouble. But as soon as the monks learned that I was a bat researcher, they asked me to investigate possible causes for their bats' decline.

The next morning, barely able to see in dawn's early light, I forced my way through dense vegetation above one of the cave's several entrances in search of a suitable vantage point to photograph the bats' return.

I was concerned about avoiding cobras in what seemed like ideal habitat, certainly not thinking about meeting other humans. But I suddenly found myself face-to-face with a surprised poacher who was obviously not happy to meet me. As we stared apprehensively at each other, I could see a fish net sagging with the weight of 30 bats and wished I hadn't gotten so far ahead of Surapon.

Despite my first apprehension and an initially tense standoff, our surprise meeting ended well. Surapon convinced the poacher that I was relatively harmless, and an excellent cash tip helped enlist his cooperation. He led us to additional bat-filled nets where we met his father, an exceptionally likable man who introduced himself as Uncle Yai. Moving from net to net, we were able to verify Uncle Yai's estimates of the numbers of bats being killed. We discovered that at least ten poachers were selling fruit and nectar bats, mostly as restaurant delicacies.

The poachers' impact was clearly substantial. On the morning of our encounter, Uncle Yai and his son alone had caught 90 Geoffroy's rousette fruit bats (*Rousettus amplexicaudatus*) and cave nectar-eating bats (*Eonycteris spelaea*), not counting the many Asian wrinkle-lipped bats (*Chaerephon plicatus*) that had been killed and discarded simply to facilitate removal from nets. At that rate, the annual take of sellable bats alone would have been unsustainable. The loss of so many bats was tragic, especially considering the ecological and economic consequences beyond the loss of bat guano sales. This trade had been growing over a period of several years, almost certainly explaining the decline in guano production.

Subsequent research at Khao Chong Pran Cave would document very conservatively that each million wrinkle-lipped bats consume nearly six tons of insects nightly, a large proportion of which are white-backed planthoppers, the dominant pest of local rice crops. Additionally, the cave's nectar-eating bats are the primary pollinators of durian, Thailand's most sought-after fruit. Each bat consumes up to twice its body weight in nectar nightly and must visit many flowers to do so.

The durian is considered the king of Southeast Asian fruits. It is football-sized with a knobby outer shell. The nineteenth-century British naturalist Alfred Russel Wallace described the flesh around several large seeds as like "a rich custard flavored with almonds." Its pungent odor evokes deep appreciation among Asians, more commonly disgust among Westerners. The decline of nectar bats threatens this key crop, not to mention other Asian products, such as petai, that also rely on bats for pollination or seed dispersal. The petai tree produces pods as long as a human forearm, with bright green beans inside. It is a popular culinary ingredient throughout most of Southeast Asia and can be eaten either raw or cooked.

As a result of our surprise meeting with the poachers at Khao Chong Pran Cave, Surapon and I were able to gather crucial data that, combined with guano harvest records from the monks, convinced the Thai government to declare the hill where the cave is located as a no-hunting zone. I explained to the monks that the economic gains from hiring a game warden would more than offset the cost of a salary. Within months, a game warden was hired, and the Khao Chong Pran Cave bats began to recover.

Prior to leaving the area, Surapon and I checked a total of ten additional caves in the region, finding that only one still sheltered even a few thousand bats. Stained limestone ceilings indicated that, in the past, several had been home to populations numbering from thousands to millions. Poles, old scaffolding, and abandoned nets clearly told the story of the bats' demise, an all too common problem throughout Southeast Asia.

Still concerned about the plight of bats in Rakang Cave, I decided to return there with Surapon to see if we could collect sufficient documentation of the economic merits of saving the cave. Certainly it was naive to allow a quarry that provided only temporary, low-income employment to destroy the long-term, more lucrative support for a whole village.

We interviewed miners and guano vendors as well as quarry workers, documenting average incomes and the number of people supported. The case for saving the cave was strong, so I prepared a special report for Dr. Lekagul, suggesting he organize a news conference at the cave. He approached both the government and news media with our findings, resulting in extensive coverage and, ultimately, a royal decree to protect the site.

Our next goal was to photographically document the importance of bats in pollinating wild bananas. For that Surapon and I drove to the Khao Yai National Park, situated in the majestic Sankamphaeng Mountains, 100 miles northeast of Bangkok.

Dr. Lekagul had obtained permission for us to work in the park, but we were in for some surprises. I wanted to get to work immediately, so I was very unhappy to be told by the park superintendent that we could not net bats the first evening. He explained that it was too dangerous without armed guards, who wouldn't be available until the next afternoon. I said, "Okay, perhaps tonight we can just net a few of the common bats here around the headquarters buildings where we'll be safe." He hesitated a bit before deciding to share some little-known problems.

At the back of the main building he pointed to the large wooden window shutters and said, "See those claw marks? They were left by a tiger that recently jumped through the window and killed one of our rangers. We don't like to draw attention to such problems, but we can't afford to have an American scientist killed. Daytime tourists rarely have problems, but being active at night can be quite dangerous." So began a week of high adventure.

The next night, Surapon and I sat beneath our mist nets, in pitch darkness, in a wild banana grove several miles down the mountain, listening intently for even the faintest of sounds. We hoped to catch a greater long-tongued nectar bat (*Macroglossus sobrinus*), the dominant pollinator of wild bananas, but we were quite nervous, hoping to net our quarry and depart as soon as possible.

We'd been assigned four armed guards, thinking we'd be protected. When we arrived at the location I'd chosen for netting, however, the guards totally refused to enter among the wild bananas. They said that tigers were bad enough, but they were not going to go where giant king cobras lurked. Our compromise was that we wouldn't tell the superintendent if they would simply agree to stand at the edge of the road while we netted some 75 feet away. We hoped to catch the bats quickly and leave prior to any trouble.

Soon Surapon and I were seated quietly, each armed with a three-foot section of metal bat netting pole, which I'd tried to assure him was sufficient for clubbing a king cobra's head in the unlikely event of one's approach. Now, every time a mouse, frog, or lizard moved, we couldn't resist switching on our headlamps. It seemed like hours but likely took only about 40 minutes to catch our first bat, a short-nosed fruit bat (*Cynopterus sphinx*). It wasn't the species we were hoping for, but we had caught it in the act of pollinating a banana flower. It was cute, and under the circumstances, we were happy to leave without our originally intended quarry.

Back at our rented cabin, we had already prepared a fresh-cut banana plant, complete with newly opening flowers. By midnight, I had the bat acclimated to captivity and to feeding from my hand and was ready to begin photographing it visiting the flowers.

By that time I was unexpectedly chilled to the bone. I had anticipated working in lowland heat and hadn't even brought a long-sleeved shirt. Fortunately, I had a small hair dryer, and about every 30 minutes I'd put it up my pants legs, then down my shirt for several minutes, or at least until I could control my shivering.

By 3 AM the bat didn't seem interested anymore and I was wondering if I'd ever get the desired photo of our bat pollinating a banana flower. Had I fed him too much? Would we have to return and catch another, more cooperative individual? I certainly hoped not. Almost an hour later, just as I was ready to give up, the bat apparently got hungry again, came to the flower, got its face completely covered in pollen, and looked up just long enough for a photo. That one picture has since been published hundreds of times in magazines, books, and a wide variety of educational materials worldwide and has had an excellent impact in helping people understand the value of bats.

There seemed never to be a dull moment during the week and a half that we worked in the Khao Yai National Park. One morning, we were awakened by screaming maids next door; they'd sighted a tiger hiding beneath the raised floor.

On another occasion, I had Surapon drop me off at an elephant trail several miles into the park, so I could find a netting site while he took our rental car to pick up the guards that were required for nighttime work. I soon found a good location overlooking a crystal-clear mountain stream that reminded me of the American Rockies except for the exotic tropical forest with its clumps of 100-foot-tall bamboo. I set three 42-foot nets along the trail and waited for Surapon to return with our guards. My goal was to catch and photograph the rather elusive lesser bamboo bat (*Tylonycteris pachypus*), a tiny species that lives in hollow bamboo stalks. It has a flattened skull, enabling it to quickly enter beetle holes an inch tall by less than a half-inch wide.

Nearly two hours after sundown, I was frankly as frightened as I've ever been. Surapon hadn't returned, and I was stranded, alone and unarmed, on an elephant trail likely to be used by tigers. Even worse, I'd accidentally captured a large nocturnal bird of the nightjar family, and while being disentangled from my net, it had uttered distress calls loud enough that I could imagine hungry tigers being attracted from all directions.

The trail I was on followed along the top of a 75-foot-high cliff. Now I was perched at the cliff edge, scanning with my headlamp for tiger eyes, fully prepared to jump off. I hoped to grab and slide down one of the huge bamboo stalks to escape. Fortunately, Surapon arrived soon after with the guards. He'd had car trouble. We never did capture a bamboo bat, something I wouldn't succeed at until many years later.

We had the misfortune to be in the Khao Yai National Park during the Christmas holidays, a time when visitors brought their own power generators and huge speakers to the nearby public campground, and they apparently came prepared to go sleepless, based on the continuous, near-deafening celebrations. Finally, desperate for sleep, Surapon and I departed for the Huay Kha Khang Wildlife Sanctuary, where we hoped to capture and photograph the greater false vampire bat (*Megaderma lyra*) and finally get some uninterrupted rest. The facilities were very nice, and since we were now located far out in pristine forest, we assumed we would find peace and quiet.

On arrival we were again escorted by armed guards, not really sure whether we were being protected from tigers, elephants, sun bears, or communist insurgents, though we were happy to have guides. Having acclimated to being cold in the mountains, we were sweating profusely as we clambered our way through dense lowland forest.

Arriving at the cave, our guards quietly inspected the entrance for signs of danger. Then I proceeded alone carrying a long-handled butterfly net and wearing a dimmed, rheostat-controlled headlamp on my head. I didn't want to use more than minimal light so as to avoid giving my quarry advance warning of my approach, but in so doing I also risked surprising other critters that could be potentially dangerous.

Passing through a narrow spot, I suddenly detected movement and realized that a six-foot cobra was attempting to exit as I entered. I froze, not wishing to make it feel threatened. Remembering a conversation I'd had with Surapon while climbing around in what appeared to be ideal cobra habitat near Khao Chong Pran Cave, I couldn't help wondering if it might be guarding eggs. Surapon had long ago reassured me that most cobras were shy and would leave me alone. Then one day we had seen a man running very fast and, relying on an apparently common Thai saying, Surapon had exclaimed, "He run like man with cobra egg!"

Still somewhat concerned about cobras, I couldn't help asking, "You mean cobras are aggressive when protecting their eggs?"

He had replied, "Oh yes." I asked at what time of year they laid eggs. His admission, "I no remember," had done nothing to relieve my current state of mind.

Fortunately, this cobra crawled on by, and I was soon able to find and catch a greater false vampire bat, which I gently placed in a nylon bag. Exiting with great caution, now with my light on full bright, I breathed a sigh of relief when I saw that the cobra had disappeared.

That evening this carnivorous bat, with its more than two-foot wingspan, turned out to be one of my favorites. I released it into the room where I was staying and, like the frog-eating bats I'd worked with in Panama, it quickly learned to come to my hand for food, in this case for gecko lizards, which it readily ate. All in one evening, I was able to obtain great photos of it catching small geckos.

Unfortunately, our arrival coincided with a special celebration for a popular man who had died of malaria 100 days before. By nightfall, mourners were expressing their grief into microphones hooked to huge loudspeakers turned to full volume. I'd never experienced anything like it before. The overloaded speakers popped and banged, sounding like shotguns firing, and when mourners intermittently tired, music continued to blare.

Surapon assured me that they'd have to quit at midnight when the station's power went off. But the mourners had apparently brought their own power generators, and the intense sound emanating just 100 feet from our bungalow rattled our windows. Even our beds vibrated. By 1 AM, I was desperately exhausted and driven nearly crazy by the intense noise, so I got into our rental car and drove nearly a mile out into the forest, hoping to find peace. No such luck! The sound remained too loud for me to sleep.

Hours later, I gave up. We packed and moved to a hotel in a nearby town. Unfortunately, our hotel was located adjacent to the town square, which turned out to be the primary gathering place for night-driving truckers, who in those days used the loudest mufflers conceivable and enjoyed honking their horns. The next morning, even more desperate for rest, we drove back to a quiet hotel across the street from Dr. Lekagul's home in Bangkok. Surapon went home, and I finally collapsed from fatigue.

Dr. Lekagul came to see me and, based on my condition, wanted to check me into a hospital. To his considerable worry, I insisted that all I needed was rest, and four days later I was able to fly home to the United States – barely.

Sadly, Dr. Lekagul passed away prior to my return nearly a decade later. He would have been proud to see all the progress resulting from his invitation and assistance. Surapon had received Bat Conservation International's first student scholarship (in partnership with the New York Zoological Society), obtained a master's degree in bat biology, and eventually became one of

Thailand's leading conservationists as secretary-general for the Wildlife Fund Thailand.

Photos from our trip together were used in a series of articles in leading Thai magazines, and the *Smithsonian* magazine featured Khao Chong Pran and Rakang Caves and the value of saving Thai bats in its January 1984 issue. My article was titled "Harmless, Highly Beneficial, Bats Still Get a Bum Rap." Accomplishments from that early trip continue to demonstrate the great value of bat conservation.

When I returned to Thailand nearly a decade later with filmmaker Dieter Plage to document the human benefits of bat conservation for the *Secret World of Bats* television documentary, we revisited both Khao Chong Pran and Rakang Caves.

On Chinese New Year, 1990, I sat enthralled watching a truly awesome spectacle: more than five million wrinkle-lipped bats emerging from Khao Chong Pran Cave only a few feet from my face. Dieter and I had just interviewed game warden Champa Molek, who'd been hired to protect the bats. And now we were documenting an amazing success.

Great columns of bats could be seen for a mile, backlit by the setting sun, which turned translucent wings bright orange. The vast numbers, flying wingtip to wingtip, nose to tail, sounded like whitewater rapids. Below, hundreds of tourists, some from as far away as Finland, watched in awe. Their tour buses were parked at the nearby temple, where local vendors sold refreshments and souvenirs.

Even before I could conduct an official count of the bats, it was clear that numbers had increased dramatically. In fact, monastery guano sale records removed all doubt. At the population low point in 1981, annual sales had fallen to 272,788 baht ($12,017 U.S.). Since Champa had been hired to protect the bats, sales had risen steadily and were anticipated to reach 2,216,500 baht ($88,660 U.S.) by year's end.

At the rate of decline seen in the late 1970s and early 1980s, even the poachers might soon have been out of work. But in 1990, they continued to ply their trade as legal hunters well away from the highly vulnerable cave entrance area where nets could not be set without catching numerous insectivorous bats that previously had been killed just to facilitate removal. At the time of my visit, the fine for catching bats in the protected area was 500 baht per bat, the equivalent of a week's pay.

Though poachers had been the original cause of bat decline, they were largely unaware of their impact. Uncle Yai was still hunting fruit- and nectar-eating bats, but at a distance, where catches were smaller and, I hoped, sustainable. Though I'd have much preferred to see an end to commercial hunting, I thoroughly enjoyed my friendship with the bat hunters. Uncle Yai's son, who had assisted Surapon and me nearly a decade earlier, had grown up and left to work in a factory, choosing not to follow in his father's footsteps.

Ood Tongcam, another bat hunter, was doing his best to support his wife and 11 children, all irresistibly friendly and considerate. Hunting bats was simply the only means he knew for supporting his family. With better education, it appeared likely that the next generation would turn to new, more respectable professions – or at least that was my hope.

Although counts of multimillions of bats in complex caves are notoriously difficult, I did my best to calculate the population size by estimating average numbers of bats per square foot and multiplying those numbers by the number of square feet of roost surface covered by bats. I also measured the areas of limestone ceilings that had been stained rusty red through long use by roosting bats. These characteristic stains often provide invaluable clues to past use by bats. Such evidence suggested the cave was capable of sheltering a substantially larger population, though I estimated the current one at close to five million wrinkle-lipped bats and 100,000 fruit and nectar bats.

Dieter, his wife, Mary, and I next drove with our interpreter to the nearby Rakang Cave. After nearly a decade I did not expect to be remembered, so I was thrilled when, moments after our arrival at the entrance, several of the guano miners shouted in recognition and came running to greet me. One young lady recounted how, as a child, she'd simply been curious about the first American ever to visit her family as I interviewed her father for my report. And Siri Tanomsri, a miner who had extracted guano from Rakang Cave for more than 40 years, told the cave's story from his experience.

The bats had started to decline more than 20 years earlier when the limestone quarry began. Although Siri believed that the blasting had disturbed some of the bats into leaving, and had killed others, he concluded that by far the greatest threat was from poaching by men who worked at the quarry. As had occurred at Khao Chong Pran, they had placed nets at the cave's entrances, catching large numbers of bats. Originally there were two major bat caves in the area, but the one farthest from the guano mining village had been more difficult to protect and thus had lost all of its bats to poaching.

The bat population at Rakang Cave was still far below former numbers. A royal decree had protected the cave itself from destruction, though limited quarrying and associated poaching had continued for six more years.

When the quarry operator finally departed, local villagers had begun guarding the cave around the clock on a voluntary basis, and Siri had been able to measure the results by the increasing amounts of guano he collected. His income had doubled in just the past year.

My next opportunity to visit Thailand came in 2012 when amateur bat enthusiast and BCI member Daniel Hargreaves and his fiancée, Heidi Cooper-Berry, offered to fund a trip for my wife, Paula, and me to photograph additional Thai bats and revisit Khao Chong Pran and Rakang Caves in exchange for teaching him how to photograph bats. It was one of the best offers I've ever had.

Our trip was organized by Jirasit Rongkaew (P'Kwang to us) and Pongsanant Trirat, who specialize in conserving bats and assisting conservation-oriented scientists and tourists. Our first destination was a site where we could photograph one of the world's most colorful bats, the rarely seen painted bat (*Kerivoula picta*).

The location was just an average rice farm except for one thing: the brothers who owned it had learned how to attract painted bats. They had planted small patches of bananas along their rice paddy dikes, and as they walked between paddies they would break the leaves about a foot from their tips, so the tips wilted and hung down in a manner that painted bats love to use as roosts. Over time, they had attracted a dozen or more pairs, and the brothers and their children knew right where to find them. The bats normally roost singly or in pairs with a single pup.

Painted bats have bodies that are only about an inch and a half long and weigh less than a United States nickel. Their fur is long and woolly and bright orange. The wings are jet black with sharply contrasting orange markings along the body and wing bones, giving rise to their name. They also are sometimes called butterfly bats, because when disturbed into flying during the day, the wings not only resemble a beautiful butterfly, but the bat has a distinctive butterfly-like flight pattern. Asleep, they are well camouflaged against the orange-brown of dead leaves.

Because of their striking beauty, these bats are often killed by collectors who mount them in glass cases and sell them to tourists. To protect the bats, our hosts keep their location secret except for a privileged few visitors who can be trusted to do no harm. Daniel had vouched for me, but even so, Pongsanant and P'Kwang were at first quite nervous about allowing me to bring one of these bats into my studio for photos.

The photo that would make the trip worthwhile would show a painted bat's beautifully marked wings in flight, but tiny bats are notoriously difficult to photograph. Since all individuals aren't equally cooperative, I'd normally hope to work with several. Given these bats' apparent fragility, though, I dared bring only one into my studio.

Just finding a suitable location had been a challenge. The only possibility was to evict the chickens from their normal roosting area beneath my hosts' home. It was built on stilts, leaving just enough space below the floor to accommodate my studio.

Daniel captured a beautiful male bat, placed it in a small cloth bag, and hand-carried it back to the studio, where I waited with a supply of mealworms. I had already arranged my camera and flashes and set up an infrared beam as a triggering device for flying bats. Now Pongsanant and P'Kwang watched from just a few feet away, ready to protest at the slightest sign of stress to the bat.

No one we knew of had ever attempted to feed a painted bat in captivity, so it was with great suspense that we all waited to see how our bat would respond. I carefully removed it from the bag and handed it a small mealworm. At first it refused. Then, on about the third try, it tentatively bit into my offering, licked its lips, reached for more, and began its first meal with gusto, much to everyone's relief.

Following a few more mealworms and a drink from a teaspoon, our bat became relaxed and was ready to be photographed. By that time, however, a thunderstorm had arrived. As lightning flashed frighteningly close, thunder roared, and heavy rain began to blow in along one side of my studio. Worried about potential damage to equipment, and concerned for our bat, we could only hope the storm would soon pass. Fortunately, despite its violence, the storm did pass within half an hour.

The first time we tried to photograph our painted bat in flight, it missed the beam, leaving us with a blank frame. The next showed a poor wing position, and so on. Many tries later, I finally had some beautiful shots, and we were able to release the bat with a full stomach near where it had been captured. I couldn't help thinking how nervous I would have been prior to the age of digital photography, which now provides instant awareness of problems and confirmation of good photos.

I spent most of the next day in a variety of awkward positions waiting for wind-blown banana leaves to pause for split seconds between gusts so I could show the tiny sleeping bats inside their leaf-tip homes. I couldn't even rely on auto focus, because this handy feature would home in on the sharply defined leaf edges instead of the softly furred bats. Daniel and Pongsanant had nearly endless patience in serving as mobile flash stands, ignoring the 100-degree heat, high humidity, and rivulets of sweat stinging their eyes.

For me, despite my enthusiasm for photographing painted bats, the trip highlight was revisiting Rakang and Khao Chong

Pran Caves 31 years after my original conservation efforts on their behalf. I had already heard glowing reports of a huge success at the latter, but nothing about Rakang since my visit with Dieter and Mary Plage in 1990, apparently in part because it had become included in a military base where few people had access.

When we attempted to approach Rakang Cave, base guards stiffly informed P'Kwang that we could not enter without a pass and that it would take a month to obtain one. Knowing how much I wanted to check on the cave, P'Kwang suggested we simply wait an hour, until the guards changed. Then we could mix with lunch-hour traffic and wave to the new guards as though we were base residents as we passed. It not only worked, but we were able to stop two officers playing golf to get directions to the cave.

When we arrived at midday, none of the guano collectors were at work, but soon one arrived on her motorbike. When we asked if she knew Siri Tanomsri, the man I'd interviewed at the cave in 1990, she smiled from ear to ear and said, "Yes. He's my father. Would you like to see him?" She jumped back on her motorbike and led us to his home. Though a bit thinner than when last I had seen him, he remained alert and in good health at 96. He no longer remembered me, though his 81-year-old wife, Jaroon, did. She had appeared in our 1990 film. The daughter said both parents had gone to the cave almost daily to extract guano into their late seventies. After all those years of close contact with hundreds of thousands of bats they remained in excellent health, not even aware of speculated disease risks.

Ten people were still collecting guano, though production had recently fallen, apparently because the now well-protected fruit- and nectar-eating bats had prospered and were crowding the smaller wrinkle-lipped bats out. Since it was only the insect-eating wrinkle-lipped bats whose guano was valued as fertilizer, thanks to its higher nitrogen content, guano mining had declined. Nevertheless, durian growers likely were prospering from increased pollination by bats without even knowing why.

In contrast, guano production had risen dramatically at Khao Chong Pran Cave. A scientific publication in 2002 reported annual guano sales of $135,000 U.S.

During our 2012 visit, Pongsanant introduced Daniel and me to the head monk at the monastery. Wisuthanathkun was an exceptionally intelligent and gracious man who enthusiastically reported on progress. He began by saying, "Of course I am very concerned about protecting the bats. Without them this monastery could not have been built." He went on to explain how he had made sure that the local community had a say in guano harvesting policies and sales, and that income was shared in order to ensure a long-term partnership in protecting the bats.

Three game wardens now worked in shifts 24 hours a day to ensure the bats' security. Also, through broad community education on the value of bats and a petition to the Thai government, Wisuthanathkun had gained legislation three years earlier that prohibited all restaurant trade in bats. He reported that stiff penalties included a 40,000 baht ($1,280 U.S.) fine or a two-year jail sentence for buying or selling bats. To the best of Wisuthanathkun's knowledge, no further bat hunting existed in the area.

Curious to double-check the monk's glowing report, we climbed to the ridgetop above the cave and checked the poachers' formerly most-used netting sites. Finding no evidence of use, we next set out to see if we could find Chamni Nakninm, former owner of the Suethanin Restaurant in the nearby town of Chong Pran. He had been the town's leading buyer and server of bats and had been very helpful to me during my original analysis of the impact of commercial trade. In fact Dieter Plage and I had filmed Chamni frying and serving bats in 1990 for our documentary.

The town had grown considerably in 30 years, so even locating the approximate site of Chamni's former restaurant was difficult. Once in the vicinity, P'Kwang got out of our van and began walking the street, asking older folks he encountered if they remembered Chamni and his restaurant. Soon he found someone who did, and we were directed to a convenience store only half a block away.

The store was Chamni's new business, and though I was now 71, and he 77, we immediately recognized one another in a wonderful reunion. He explained that he'd had his restaurant until three years earlier, when restaurant trade in bats had been outlawed. He had specialized in serving fried bats and cane rats, but the new law had put him out of business, so he'd converted his restaurant into a convenience store.

Chamni had seen *The Secret World of Bats* and was well aware of my role as a bat conservationist, but that didn't seem to dampen his enthusiasm at seeing me again. To his knowledge, no one sold bats in restaurants anymore. He explained, "Before hunting and selling bats was outlawed, I could purchase bats for 10 baht apiece. But now, even if I wanted to risk being arrested for selling bats, I would have to buy them for at least 100 to 150 baht each, which would be too much for me to make a profit."

That evening, a Sunday, we returned to Khao Chong Pran Cave to see for ourselves the huge tourist trade the bats had attracted. Several hours before the bats' emergence, street vendors began setting up their booths below the bats' flight path across the valley. By the time the bats appeared, we estimated there were several hundred vendors and 1,000 to 2,000 tourists.

Vendors were selling almost everything imaginable, from bat dolls, T-shirts, and food to furniture. The bats had become very big business, and I couldn't have been happier. I could only hope that my former poacher friends, who could no longer be found, had discovered a way to share in the prosperity.

Just as this book was going to press a team of American and Thai scientists published their findings conservatively estimating that Khao Chong Pran bats are now saving local rice farmers more than $300,000 U.S. annually in avoided crop damage from white-backed planthoppers. Clearly, bat conservation is paying big dividends!

CHAPTER 12

MYSTERIES OF BAT-GUIDING FLOWERS

WE WATCHED QUIETLY beside a rainforest trail as peccaries passed and mosquitoes swarmed. We had set three mist nets in hopes of catching brown long-tongued bats (*Glossophaga commissarisi*), the area's only known pollinators of the sea bean (*Mucuna holtonii*).

I had teamed up with German colleague Ralph Simon, an expert on how highly specialized flowers use echolocation guides to lure bats, and we had successfully proposed a story to *National Geographic*. Now we were spending our first evening together in the field, accompanied by my wife, Paula Tuttle, and Susan McGrath, a professional science writer assigned to the project by *National Geographic*. We had come to the La Selva Biological Research Station in Costa Rica to take the first photos of a sea bean flower firing pollen onto a bat.

This incredibly specialized plant relies exclusively on bats, avoiding bees and hummingbirds by waiting until it's fully dark to open its flowers. It then provides just one tiny opening that can be found only through the use of echolocation. A reflective petal rises, signifying readiness. The petal also serves as an echo reflector that guides approaching bats like landing lights on airport runways. When the bat's long tongue enters the "keyhole" to obtain a sip of nectar, it triggers spring-loaded anthers to fire pollen. The whole process can be completed in a fraction of a second.

Having found some flowers over the trail, we had set our nets at sundown, unaware that the bats wouldn't arrive for another hour. In the meantime, we tried not to frighten potentially approaching bats with all our mosquito swatting.

We had already set up my photographic studio in the cottage we had been assigned. We had promised Kathy Moran, *National Geographic*'s senior editor for natural history, that we could get a series of shots showing a bat sending high-frequency echolocation sounds into a reflective petal, followed by its tongue entering the flower's triggering point and the flower firing pollen onto the bat's rump. We knew this would take lots of time. First we needed bats. As the new one struck our net Ralph and I rushed to get ahold of it. We breathed a big sigh of relief when we recognized the species we had come for. Then Ralph groaned, "Oh no!" Its breasts were full of milk. We'd caught a nursing mother and would have to release her. After catching and releasing two more nursing mothers, Paula asked, "What if they all have pups?"

The next two bats we caught were small fruit-eaters, important to the rainforest ecosystem but not the ones we needed. When Ralph checked the third net, he whooped with joy as he announced, "We've finally got a male." Two females without pups followed, so after an hour and a half of mosquito misery, we closed our nets and returned to the studio.

Ralph prepared a mixture of honey and water, and we fed each bat by hand prior to its release into the studio, where we had also filled a hummingbird feeder. Within minutes they had adapted to the low light and were coming to the feeder. One even learned to follow Ralph around for free drinks.

Torrential rains poured for the remainder of the night, and by the following morning the nearby Sarapiquí River had overflowed its banks. The trail where we had caught our bats was flooded beneath six feet of water, and all the reserve's trails were closed. We were very lucky to have already caught our most important bats. But the problem of reaching flowers remained.

These flowers grow in circular clumps that botanists refer to as *inflorescences*. Each plant produces multiple inflorescences that hang several feet below surrounding vegetation in a manner that facilitates bat discovery and visitation. A series of flowers opens at intervals through the night so feeding bats are forced to return again and again to meet their energy needs.

The following afternoon, while I set up a camera, flashes, and other equipment in the studio, Ralph, Paula, and Susan went out in search of flowering sea bean plants. Because these plants grow mostly along rivers that were now flooded, finding some with reachable buds was a challenge. Nevertheless, by midafternoon they had found some they could reach with the aid of an extension ladder and long, extendable pole clippers. They returned proudly bearing inflorescences, one of which Ralph and I hung in a natural-looking set in my studio.

That evening we waited with anticipation, but the buds never opened. Disappointed, we realized we had picked them too early. Thereafter, we would wait until after sundown to cut their long stalks, though climbing at night, often in the rain, was risky.

The next day, our determined crew located a new sea bean vine. This time they waited until after dark to cut the inflorescences after some flowers were beginning to open. As a further precaution we immediately immersed the stems in bottles of water to keep them fresh. Several inflorescences were needed, because each flower would fire pollen only one time. Thereafter, it would be useless for our purposes. We hung the extra inflorescences under a nearby staircase outside the studio, so we could present them to our bats one at a time as the night progressed.

Looking at the first six-inch-diameter inflorescence we had just hung, with its multiple flowers, I realized that I could frame and focus on only one at a time. How would I know which one a bat would visit first or when a bat was merely inspecting? And what would I do if multiple bats approached at the same time? These were just some of the questions on my mind as we expectantly waited for our bats to become active.

When the first bat arrived, virtually all my worries were verified. All three checked out our flowers at the same time. The first and second flowers they visited were not the ones I had focused on, and when they finally visited that one, everything happened so fast that, even shooting at four frames per second, I was not nearly fast enough to catch the pollen being fired. In a matter of two minutes, three flowers had been triggered without my getting a single useful photograph. Furthermore, I noticed that, like guns, some were hair-triggered while others weren't.

Quite aside from scouring the forest daily for new buds, we had to learn our bats' behavior, how to anticipate their responses to different set arrangements, and how to predict and guide their final approaches. A bat's first approach was almost always exploratory, the second being the one to photograph. Also, one of our first three bats turned out to be a relatively unsophisticated youngster who too often missed the target and wasted our time and flowers. That one had to be released.

By manipulating the positioning of plant material and flash stands we could influence the angle and direction of a bat's approach, and through careful aiming of a dim lamp I could see an impending visitor earlier. This enabled me to be better prepared to time my shooting. At the beginning I assumed that shooting at four frames per second during an approach would be helpful, but I was mistaken. A bat's final approach, triggering, and pollen firing could occur in such quick succession that, even at four shots per second, starting a sequence just a tiny fraction of a second too soon or too late could result in missing all three stages. Shooting just one shot per approach proved more successful.

Even when my timing was perfect, there was always a chance that a wing in the wrong position would hide a critical event. Thousands of shots were almost, but not quite, right, explaining why some 10,000 images were required before at last our exhausted team could pack and head for home.

Our next priority was to photograph pollination of the echo vine (*Marcgravia evenia*), a rare plant found only in Cuba. No one seemed to know much about how to go there legally, with regard to either United States or Cuban customs, though it seemed that people did it all the time. Most managed it through religious or cultural exchange programs, aided by experienced professionals. They warned that a "do-it-yourself" approach was risky because of the severity of penalties for noncompliance with United States prohibitions, including fines of $10,000 or more.

After many failed calls to the United States State Department, I finally reached a secretary who said I would need Treasury Department clearance. She forwarded me to their Office of Foreign Asset Control (OFAC), who finally emailed me the regulations. The document was formidable, but for my situation it was simple. Though I had been told repeatedly that I would need to have a General License, which sounded complicated to obtain, I now read that I just needed to carry a vita, showing me to be a full-time professional, and an outline of my work plan in Cuba. It even said, "No prior written approval from OFAC is required for travel under a General License."

The next unexpected hurdle was getting an okay for travel to Cuba from *National Geographic*'s travel department. They simply couldn't believe that any United States government license could be so simple. To convince them, I had to ask my congressman's office to provide a letter of confirmation.

There was one more stateside hurdle. My wife, Paula, had become an integral part of our team, and she was determined not to be left at home. She couldn't pass as a full-time professional; however, after further research, I learned that United States sanctions against travel to Cuba did not actually prohibit travel there. American citizens simply weren't allowed to spend any money in Cuba. To avoid that problem, our German colleague, Ralph Simon, agreed to cover all Paula's Cuba-related expenses and provide her a letter on his university letterhead offering her an all-expenses-paid volunteer opportunity. She could carry his letter in case of challenge by United States customs.

Greatly relieved about having finally solved all our stateside problems, we now had to face the Cuban government. Ralph had minimal problems because he was German and had held previous work visas for his graduate research there. We were also assured by our Cuban colleagues that Paula could obtain a tourist visa at the airport in Havana and would be permitted to travel with us to most places without a problem.

I, however, still had a problem. There was no escaping the fact that I was entering to work. My 350 pounds of photographic equipment would be a dead giveaway. With help from Ralph, we applied four months early for a work permit for me to help him photographically document his research discoveries.

Unfortunately, the Cuban official who had graciously agreed to walk my paperwork through the process was involved in an accident that required hospitalization, and in the ensuing confusion someone apparently forgot about my application for a work visa. As I became increasingly concerned, we delayed our planned departure by a month. Then just five days prior to leaving, with tickets already in hand, we were warned that my work visa had not yet been approved.

Ralph was already in Mexico on tickets that could not be altered, so *National Geographic* agreed to cover his expenses to continue on to Havana and Santiago de Cuba, 540 miles southeast of Havana, to meet with our intended hosts at the Centro Oriental de Ecosistemas y Biodiversidad (BIOECO) regarding plans to postpone our trip for a year. He and their staff would also confirm locations and flowering times for echo vines. At least theoretically, next time all would go smoothly.

A year later, in March 2012, we entered Cuban customs in Havana with substantial trepidation. Our BIOECO hosts had emailed us that visas would be ready at customs, but that didn't stop us from worrying about possible lapses. We'd also been warned about potential problems clearing customs with so much valuable equipment.

A lot of what we had heard about Cuba turned out to be simply untrue. We had read warnings not to get sick in Cuba for fear of poor medical care. But we had to purchase Cuban health insurance on arrival because so many sick foreigners had been coming to Cuba for treatment. And the dreaded customs agents had all our paperwork in good order and couldn't have been more courteous and helpful. They did meticulously check my gear, though I suspect the inspection took longer than necessary because of the agents' fascination with my tales of training and photographing bats.

On exiting, we were met by Carlos Mancina, one of Cuba's leading bat researchers, who would join us later in the trip, and our BIOECO biologist-guide, Angel Eduardo Reyes Vazquez, who was introduced to us as Yayo. Yayo would accompany us on our flight to Santiago de Cuba and by rental car for the rest of our trip. He was built like one who could win an Iron Man contest, and his extra strength and endurance turned out to be indispensable when it came to hauling heavy photo gear up treacherously sharp limestone cliffs to caves.

The next morning, joined by a second BIOECO biologist, Margarita Sánchez Losada, we set out in two small rental cars for our first field site, about 100 miles to the northeast. Crammed in among piles of field gear, I drove one car, Ralph the other. I was unusually nervous driving on narrow, curvy roads cluttered with more horse- and ox-drawn carts than automobiles. We had heard vague rumors, probably exaggerated, of tourists having been sentenced to life in prison for causing accidents.

As we passed the infamous Guantánamo Bay, we were stopped at our first police roadblock, fully expecting to have to show our documents and submit to some kind of search, but were surprised when we were treated almost as if we were foreign dignitaries. Only our Cuban guides had to show identification and answer questions.

We arrived at our first work site, the Campismo el Yunque, a half-hour's drive from Baracoa, by late afternoon. The manager proudly welcomed us as his first foreign guests. With his help we soon had our photo studio set up, and by sundown had set bat nets in front of several kinds of bat-pollinated trees. Within an hour we had caught two Leach's long-tongued bats (*Monophyllus redmani*) and a Cuban flower bat (*Phyllonycteris poeyi*), a species endemic to Cuba. We hand-fed them our standard honey-and-water mixture, made sure they had learned to feed at our hummingbird feeder, and turned in early, meaning by midnight.

The next day we all joined in the hunt for the rare echo vine. We scoured the surrounding countryside, from river bottoms to mountaintops, to no avail. There were simply none flowering.

That evening Ralph and I built a nice set featuring the flower of a blue mahoe tree (*Talipariti elatum*), and Paula and I had to wait until nearly 4:00 AM before our Cuban flower bat finally got hungry enough to visit for a sip of nectar. The resulting shot of the bat as it emerged from a bright orange hibiscus-like flower, completely covered in pollen, was worth every frustrating hour of our vigil. It well illustrated why plants compete for bats as primary pollinators, as well as demonstrating their importance to human economies. The blue mahoe is a valued timber tree in Cuba.

We were excited to finally be in the field with our Cuban colleagues, getting some good photos. Nevertheless, more frustrating surprises awaited. When Ralph, Margarita, and Yayo returned that afternoon from an expanded echo vine search, I could see the discouragement in their faces. Ralph reported, "We have searched everywhere we can think of and can't find a single plant that is even close to flowering."

Much more calmly than I felt, I asked, "Do you have any other ideas?"

His response, "We know of another possible location, but it's a six-hour round trip to go there. Yayo and I can check it out tomorrow." We had just gotten the studio organized with acclimated bats, ready to work, so I hated the idea of a possible move, but I agreed we should see what they could find.

They left early the following morning, while Margarita waited with Paula and me. Ralph didn't call until after noon the next day.

"We've found them," he said. "Several, and they're in full bloom. We've cut two inflorescences and are bringing them back with plenty of background material." While relieved they'd found some echo vines, I groaned inwardly, fearing the leaves would wilt in the hot weather prior to arriving back at the studio. They got back shortly after dark, and Paula and I were extremely relieved to see that Yayo had managed to keep the stalks submerged in bottles of water despite hours of bumpy, twisty roads, and the leaves hadn't wilted.

As quickly as possible we entered the studio, and Ralph and I began building a natural-looking set from the vines he and Yayo had brought. In the meantime, Paula guarded the first inflorescence against premature visitation by our bats. Her hands had to be constantly on two sides of the inflorescence as she alertly kept an eye on the bats. Every time they'd begin to fly, I'd yell, "Watch out! One's coming in from behind you. Keep your hands moving. Don't touch the flowers!"

The flowers were red and had long been thought to be pollinated by hummingbirds. This was a logical assumption, since many bird-pollinated flowers are red and because hummingbirds routinely visited these by day. Nevertheless, when Ralph and his major professor, Otto von Helversen, first saw that echo vines had an upturned, dish-shaped leaf above each inflorescence, they hypothesized these to be echolocation reflectors designed to guide bats.

When they used a low-light camcorder and infrared light to record at night, they saw Leach's long-tongued bats visiting the flowers and hypothesized them to be the primary pollinators. Ralph's later lab experiments, combined with ultrasonic recordings and additional observations of both bats and hummingbirds visiting flowers, would validate that bats are the primary pollinators and that the dish-shaped leaves indeed provide echo signals to guide bats.

When we finally lowered the lights and sat still, one of our bats immediately approached, flew away, and then returned to more closely examine our offering. Clearly, it was interested, but would it finally land from the right direction, or would it come in with its back to the camera?

Paula knew the routine. I pointed and whispered, "Please stand right there." I wanted her to interfere with the bat's ability to approach from the wrong side. If it landed wrong, I'd get no more than a boring butt-shot while it knocked off pollen needed for an ideal photo.

Margarita and Yayo watched silently outside the studio, and we all held our breath in suspense as we waited to see what would happen. Fortunately, the bat performed perfectly, approaching exactly where needed, and I clicked off four frames in the first second. We shot almost continuously for the next half-hour, then rearranged the set with the second inflorescence.

Each inflorescence consisted of 20 to 25 upside-down maroon flowers, arranged in a horizontal whirl, with several anthers each, full of yellow pollen, hanging below. About an inch and a half beneath the anthers hung a circular cluster of bright red nectaries (nectar-filled containers) resembling Grecian urns.

We had nailed the needed shot of a bat completely filling the gap between nectaries and floral reproductive organs in a manner that transferred maximum pollen. Now we needed photos of hummingbirds feeding, illustrating how their smaller heads fail to fill the gap as required to achieve pollination.

The hummingbird shots would require working in the wild, and we went to bed at 3:30 AM, planning to pack up and move to the new location later in the morning. By 7:30 AM Paula and I were up, getting ready to move. We were all in an upbeat mood at breakfast. Then Ralph drove Margarita to a government office in Baracoa, "just as a formality," we were told, to let them know we were moving. But, to our utter shock, she was informed that yes, our permits allowed us to go there, but we couldn't take pictures. That killed our entire plan. Ralph and I were in complete disbelief. As far as we knew, this was our only hope of obtaining the final images that *National Geographic* had requested.

Then Ralph had another idea. Might we find echo vines in bloom if we drove to the Humboldt National Park and hiked up into a ridgetop cloud forest where the climate would be different? It would be a two-hour drive to the park, then a 45-minute hike up to the cloud forest. The coastal highway to the park could easily be in *Guinness World Records* for roughest highway. It was so bad that local Cubans called it an extinct highway.

The next morning we made the grueling drive, hired a park guide as required, and hiked up the mountain. We found several flowering echo vines, but only one was reachable for photography, and it would be a long way to carry tripods, flash stands, and other heavy equipment up slippery wet trails.

Luckily, Yayo and our local guide didn't mind showing off their superior strength, so the next morning we returned with them in the lead with the heaviest equipment. Just as we started up the worst part of the trail, the sky burst in a deluge that made umbrellas and raincoats irrelevant.

Thirty minutes later we arrived at our chosen site, nearly hypothermic from the wet cold. From the looks of the dark clouds, it appeared the rain would last all day. We just stood there in bedraggled misery. But 20 minutes later, the rain suddenly stopped and the sun popped out.

We did our best to shake the water from dripping trees overhead and arranged my high-speed flashes on stands. Twenty feet from the inflorescence, I lay back with my head propped against a sturdy equipment case. My camera and telephoto lens were mounted on a tripod in front of me, and I reclined, with my right knee raised. As I clutched a cable release firmly in my right hand, just inches from my chin, I suddenly felt an anole lizard climbing my raised leg. And so began the equivalent of a reptilian soap opera right in front of my eyes.

Since I was perfectly still, waiting for a hummingbird to arrive, the climbing anole apparently mistook my knee for a knot on a log and selected it as a place to court. His display consisted of head bobbing, combined with periodic extension of his dewlap – a bright orange flap of skin beneath the lower jaw. A short time later, I noticed an approaching female. She jumped from a nearby branch to my left arm and proceeded across my chest and up my right thigh to the male.

But, just as the lucky guy was claiming his prize, I felt another anole climbing my leg. It was a second female who objected. She attacked the pair, and both females promptly abandoned the male, whereupon he noticed his reflection in my glasses and, assuming it to be a competitor, scrambled up my arm and attacked the nearest lens. At that point I couldn't resist laughing. Surprised, he fled. Sometimes courtship just doesn't go according to plan!

Nearly an hour later, the first Cuban emerald hummingbird approached, spotted my setup, and quickly left. Minutes later it was back. I got one photo, but the flashes scared it away. They had to be quite close to balance with sunlight. By 4:00 PM, the hummingbirds had become accustomed to me and my flashes, and I was getting hundreds of excellent shots. From the photos, it became abundantly clear that hummingbirds were thieves, not pollinators. They consistently sipped nectar without touching floral reproductive organs.

On the drive back that evening, I reminisced about a similar experience I'd had 30 years earlier in Kenya. I had noticed that the showy white flowers of giant baobab trees opened after sunset and dropped to the ground the next morning. When I questioned a local wildlife expert, I was assured they were pollinated by bush babies – small, nocturnal primates that often live in hollow baobabs.

The next evening I went out with my headlamp and immediately spotted a bush baby sipping nectar from an open baobab flower, but it was reaching down from above in a manner that entirely missed reproductive organs, which hung several inches below the petals. Later I would observe nectar-feeding bats swooping up from below in a manner that fully picked up and transferred pollen. Also, a bush baby could find all the nectar it needed without moving to another tree. In contrast, the bats arrived in groups that quickly moved from tree to tree and could cover long distances. Throughout my career I've often been surprised at the frequency with which bat contributions to nature and human economies have gone unrecognized. In fact, most of the world's more than 1,300 bat species remain unstudied.

Prior to returning home we had one last stop in Cuba, the incredible Cueva de los Majaes – in English, Cave of the Boas. This several-mile-long cave is located high on a rugged limestone ridge, protected by the Siboney-Jutici Ecological Reserve.

Led by station director Jorge Tamayo and fellow bat researcher Carlos Mancina, we had climbed from ledge to ledge up razor-sharp limestone cliffs to reach the entrance. We were sweating profusely as we finally set our cases of equipment just outside. It would take at least three hours to set up a system for photographing emerging flower bats. First we had to carefully position an infrared beam where we anticipated emerging bats would pass. The beam was connected to a radio transmitter that would communicate with additional transmitters attached to each of three flashes, all of which would fire at about a 40,000th of a second each time a flying bat passed through the beam. In the sharply sloping, rugged terrain, just securely positioning two stands for the beam and three more for the flashes, plus a tripod for the camera, was a time-consuming challenge.

At sundown, we waited quietly for the bats to emerge, joined by several three- to nine-foot-long boas and dozens of large land crabs, all waiting for the bats. We had to remain extraordinarily quiet as these bats were extremely wary, a necessity when so many predators were waiting.

Soon the flashes began popping, though only a small fraction of shots would be great. I was perched on a ledge with my camera and tripod about 12 feet above the floor, trying to be perfectly still, when I felt something nudging me just above the belt line on my back. I didn't dare move or turn on my headlamp to look, for fear of scaring the bats, so I just waited, knowing it couldn't be anything worse than a crab or a boa. A couple of minutes later, just as I was becoming accustomed to the probing, I couldn't help yelling in response to a sudden stabbing pain. My light revealed a six-inch-diameter land crab biting my back.

As we waited for the emergence to resume, I wondered how many flowers a million nectar bats would visit that night and how far they'd have to travel to find them all. Answers are not yet available, but when they are, convincing people to protect major bat caves will be a whole lot easier. This one cave was estimated to shelter several million bats of a dozen species.

The following morning, with our Cuban work complete, Paula and I packed, bade fond farewells to our Cuban colleagues, and headed for Havana.

Entering customs in Houston was far simpler than we'd expected. All prepared with my documents, I announced to the agent, "I'm just returning from a research project in Cuba." He smiled, asked, "How'd you like it?" and stamped my passport. After all my early agonizing, it was hard to believe I'd just been cleared without so much as a glance at my documentation.

Immediately on arrival home, we had to begin preparations for our final quest, a rare cactus found only in one Andean valley in Ecuador. Years earlier, while working on his Ph.D. research, Ralph and his major professor had noticed a cactus whose flowers were surrounded by a soft, beardlike material. Since this plant's flowers produced copious nectar and pollen, and opened only at night, it was nearly certain to be bat-pollinated. But, additionally, he hypothesized that the soft fibers surrounding the flowers might be assisting in echolocation by playing an opposite role from the flower petal and leaf reflectors we had documented in Costa Rica and Cuba – in other words, absorbing rather than reflecting bat echolocation signals.

This plant was known locally as the old man's beard cactus. Scientifically, its name was *Espostoa frutescens*. Ralph had expected it to bloom in January or February, based on previous reports, but it still hadn't flowered by late April. Then our Ecuadorian colleague Vinicio Santillan emailed us that he had found several plants in bloom on the first of May.

Paula and I packed in great haste and took off for Susudel, a small town high above the Rio León, whose valley supported old man's beard cactus. Vinicio had reserved several rooms for us at a wonderful bed and breakfast at the Ferbola Organic Farm. We unpacked and set up the studio that evening, and the next morning we drove into the valley in search of the cactus.

To Vinicio's dismay, the cactus plants he had counted on had just been plowed under by a new landowner who was planting corn. So he took us to an alternative site less than a mile away. We found cacti that had bloomed as recently as the previous night, but no buds remained. That evening, Vinicio had to return to Cuenca, but we learned of a local shaman, Juan Valdi, who was reputed to know a great deal about local plants. For the next two days, Paula and I traveled some of the most rugged terrain imaginable with Juan. We drove along roads cut in the sides of sheer precipices and crossed bridges that consisted of nothing more than eucalyptus poles laid side by side across gaps. One broken pole, and we could have plunged thousands of feet to the river below.

Even the hiking was scary. Unconcerned about the dizzying heights, Juan led us down barely visible trails where a single misstep could have meant a fall of hundreds of feet. We sometimes descended as much as 1,500 feet, only to climb out again empty-handed. When we would finally reach less precipitous terrain, we'd barely be able to inch our way through tangles of some of the thorniest plants I've ever encountered. Yet there were also pleasant surprises. By simply crossing a ridge we would, in just a few feet, pass from desert to lush, flowering cloud forest.

The next day we were joined by Nery Fabian Chamorro Rodas, a veterinarian who had become interested in bats through a workshop I had taught several years earlier in Paraguay. His enthusiasm and climbing skills were invaluable. Still aided by the shaman, we tried a new location a two-hour drive up the valley. Late that afternoon, Nery finally spotted a cactus bud.

We carefully cut the branch, impaled it on a pointed aluminum rod, and carried it back to our van. We had no idea how long it would be before it opened but were excited to finally have at least one bud. En route home we ran into a section of road that had become a quagmire from an unusual storm and we got badly stuck. Juan warned that it would be dangerous to still be there after dark, as this was a drug-running route. With the help of nothing more than a machete and a short section of bat net pole, we got ourselves dug out just as the sun was setting.

We now had two concerns: avoiding contact with the drug runners and hoping our bud wouldn't open that evening. Hours later, we breathed a collective sigh of relief as we arrived back at the Ferbola Farm with our bud still closed. But even if our bud opened the next night, we still faced a big challenge. We didn't yet have any bats, and the windswept mountainsides didn't afford much opportunity for successful bat netting.

The next morning we drove down to the much warmer valley floor where I had spotted a farm with several banana plants. Through long experience I had learned that, worldwide, there is seldom a more certain magnet for nectar-eating bats than banana flowers.

All the bananas we so love to eat originated from wild, bat-pollinated ancestors in the Old World tropics. But wherever humans have imported banana plants, nectar-feeding bats have found them. I hoped this would hold true in the Andes.

The farmer, Luis Salazar, took an immediate interest in our work and eagerly encouraged us to set our nets among his plants.

As dusk settled, we didn't have long to wait. Within the first 30 minutes, we caught two common long-tongued bats (*Glossophaga soricina;* also called Pallas's long-tongued bat) and quickly returned to the studio with our precious cargo. As usual, we hand-fed them and watched to be sure they were eating from the feeder. We still had no idea whether or not they would visit our cactus even if its bud finally opened. After all, we had captured only one of several nectar-feeding bat species that might pollinate this particular cactus.

Paula and I were awakened early the next morning by Nery pounding on our door. Clearly alarmed, he reported, "*Los murciélagos están muertos!*" The bats had been doing so well when we'd left them, it was hard to believe they could be dead. Sure enough, when I arrived, both were lying on the floor. One was indeed dead, but the other, though cold to the touch, was still breathing.

As I had done many years earlier in Africa, I quickly warmed the survivor and called for honey water. Within minutes, the bat was eagerly lapping, and I knew it would be fine. But what could have caused this problem? We needed an answer fast.

Since such bats routinely fed at much higher elevations in the Andes, it was initially hard to believe that temperatures in the 50s were too low for their survival. Nevertheless, these bats likely lived near where we had caught them, more than 1,500 feet lower in altitude in a warmer climate. They probably lived in a heat-trapping roost and shared body heat with others when inactive.

Finally realizing the problem, we borrowed a butane stove, and for the remainder of our trip, we heated the bat room 24 hours a day.

This unfortunate experience raised a fascinating question. How could plants growing above timberline, where the climate was too cold for bats to live, afford to rely on bats as primary pollinators? Nectar-feeding bats apparently lived in subtropical valleys from which they traveled nightly to reach rich food sources thousands of feet above. The question of how bats can afford such arduous climbs remains unanswered, but I suspect that, like the famous Andean condor, nectar bats may ride thermals, which could make such trips far less costly.

The following morning, Ralph Simon arrived from Germany. That evening, acutely aware of the precarious situation we were in with regard to having the right captive bats, we returned with Ralph to Luis's farm. We needed a minimum of two species of nectar-feeding bats prior to obtaining more cactus blossoms in case one didn't like this particular plant. Based on our Costa Rican experience with bats having specific preferences, it was hard to be confident. By netting until later in the evening, we managed to capture one more common long-tongued bat and two Geoffroy's hairy-legged bats (*Anoura geoffroyi*).

While waiting for bats to get caught, we explained our problems finding blooming cactus plants to Luis. Local farmers had cultivated formerly prime habitats, and ranchers had set fires that had effectively killed nearly all the cacti except those protected by rugged cliffs above the Rio León. It had also become apparent that no one knew for sure what triggered flowering. The shaman, Juan, thought it was the moon. Luis suspected it was rain.

Though anxious to help, he warned, "Last week's rain was probably the last we'll have for at least another three months." The usual dry season was beginning.

That night we worked especially hard to be sure our bats were well acclimated.

At nearly 11:00 PM Nery came bursting through the door to the room where Ralph, Paula, and I were working with the bats, exclaiming, "The cactus flower is opening!" Greatly surprised, we all ran outside to see. Sure enough, it was already two-thirds of the way open.

Quickly, we began preparing a set and brought the flowering branch inside. It was almost completely open as I turned a bright headlamp on it to frame and focus my camera. Then we noticed it was starting to close, apparently tricked by my headlamp into thinking it was morning. Fortunately, we were nearly ready, so we were able to quickly dim the lights.

Now we waited in absolute silence, desperately hoping that one of our still not-well-tamed bats would approach our only cactus flower before it closed. More than an hour passed, and not one bat moved. We continued to wait, fearful that we had the wrong bats, or that they might still be too afraid of us. No one had ever seen a bat visit an old man's beard cactus, so we had no idea what to expect.

At nearly 3:00 AM our common long-tongued bats began to circle the studio in a flight pattern indicative of searching for food. Then one of them swooped up in front of the cactus's bearded material, apparently sweeping it with ultrasound. Moments later it returned and thrust its head deep into the flower, hovering there for less than a second before departing. When the same bat returned, head covered in pollen, I clicked away, obtaining several nice shots, but not the perfect one I'd want to show to *National Geographic*. The bat wasn't ideally positioned relative to the flower.

Later that morning, we were surprised to hear it raining. Perhaps if Luis was correct, we would now find more buds. Luis led us to several new cactus locations, still exceedingly rugged but accessible, and we were relieved when Ralph and Nery did find several new buds. Thinking about the places they had to climb gave me nightmares of broken necks and worse, but, fortunately, no one fell.

Over the next week it rained daily, we took hundreds of new pictures, and Ralph was able to use the studio to make the first ultrasonic recordings to document that the soft, beardlike material actually absorbs bat echolocation signals in a manner that contrastingly highlights echoes from the bowl-shaped flower.

In this case both bat species came to the cactus. In fact it was one of our Geoffroy's hairy-legged bats that ultimately provided the image that appeared in the March 2014 issue of *National Geographic.*

Why do plants go to such great lengths to attract bats? As Ralph would say, "Bats simply carry more pollen farther." Through association with Ralph, I've learned that the diversity and abundance of bat-pollinated flowers is far greater than even I had ever imagined. From the lowland rainforest of Costa Rica to the Andean paramo of Ecuador, I was amazed at the number of ecologically essential plants that are tuned in to bats. One could spend an entire lifetime studying their co-evolved sophistications in a single location without running out of new discoveries.

CHAPTER 13

BAT FORESTERS

LOCKED IN AN UPSTAIRS ROOM of a frontier brothel, I carefully barricaded the door with my 350 pounds of luggage. Just moments before, I'd hastily been given refuge by one of the proprietors, who had rescued me from a drunken brawl. I still felt far from safe, having no idea what might happen next.

I had paid an exorbitant price for a taxi driver who had claimed to know the way from Abidjan to the Lamto Ecological Research Station, 100 miles to the northwest. But by two in the morning, it had become obvious that he was lost. With him speaking only French and me only English, there was little hope of negotiation. He simply pulled into a seedy-looking bar, filled with drunken loggers and prostitutes, spoke to an apparent owner, and began unloading my equipment.

As far as I could tell, the establishment was owned by two Lebanese brothers who had immigrated to the Ivory Coast. Thanks to hopeless language barriers, I never did get the story straight. But one brother wanted to help rescue me, while the other unmistakably hated Americans. He began yelling epithets and calling me some apparently really nasty names, whereupon several drunk truckers jumped up in my defense. Seconds later all hell broke loose as a full-blown brawl erupted.

In the nick of time, the first brother managed to recruit half a dozen helpers who each grabbed a piece of my luggage and started scrambling up the stairs. I had little choice but to follow.

The trouble had begun when my bat research colleague, Don Thomas, on his way to meet me at the Port Bouet Airport in Abidjan, had been robbed. While attempting to aid police in apprehending the culprits, he had missed meeting my flight.

Then, in the next of an unfortunate cascade of events, customs officials had decided to confiscate all my photographic equipment. Not daring to let such expensive gear out of my sight, I had attempted to go with it, and they had adamantly refused. Just as we were at a standoff, with them threatening to arrest me, an American diplomat, whom I'd met on the flight from Nairobi, came to my rescue.

A heated argument ensued, and my equipment was finally released. By that time, Don had given up hope of finding me. (This was long before the advent of cell phones.) Airport moneychangers and banks were closed. I couldn't cash travelers' checks to hire a taxi, so my new diplomat friend offered me a ride to his hotel. He convinced the hotel to cash just enough to pay my cab fare and negotiated with a driver to take me to the research station. Though this seemed reasonable at the time, it had turned into a terrible mistake.

After spending a sleepless night barricaded in my room, I welcomed a knock on my door the next morning. It was my rescuer of the night before. He was very nice, made sure I had at least a meager breakfast, and took me to a tiny, makeshift bank to change some travelers' checks. When the bank adamantly refused, he took me to a one-room school, hoping the teacher could speak some English so we could get beyond mere gestures and facial expressions for communication.

The teacher's English amounted to no more than a few broken phrases, but somehow I finally communicated that I needed to get to the Lamto Ecological Research Station and that I would be happy to reward my rescuer generously if he could take me there.

It was a long drive, but we made it by midday. Don was relieved to see me and helped cash some travelers' checks to reward my rescuer. He also translated my deepest appreciation into French.

The research station where he was working was situated in a savannah studded with tall palms and low, shrubby trees. It was bordered by an extensive forest on one side and by a narrow strip of streamside gallery forest on the other.

I had come to help document Don's pioneering research on the role of fruit-eating bats in reforestation. In those days, nearly all seed-dispersal studies were limited to birds, making his work especially important.

For his Ph.D. research, Don had studied the impact of bats as seed dispersers in African savannahs. For more than a year he had been setting his mist nets to capture bats as they foraged in varied forest and savannah habitats. He had learned that straw-colored fruit bats (*Eidolon helvum*) were migrants that arrived from the coastal forest zone into the savannah area at the beginning of the rainy season, retreating back several months later as the rains ended. Other species, such as the Buettikofer's and dwarf epauletted fruit bats (*Epomops buettikoferi* and *Micropteropus pusillus*) and Angolan fruit bats (*Lissonycteris angolensis*), were year-round savannah residents.

His results were providing the first clear documentation that both migrants and residents had an important ecosystem impact. Straw-colored fruit bats were the primary seed dispersers of huge iroko trees and served as equally important pollinators of the kapok tree, another forest giant. Resident bats relied year-round on smaller fig tree species that were the first to reclaim savannahs.

In temperate zones, forests tend to be dominated by trees that produce small, dry seeds that simply fall to the ground or are scattered by wind. This works fine where summers are short and seed-eating insects don't have time to produce large populations between winters. In tropical climates, however, very few such seeds escape long enough to germinate, probably explaining why most tropical plants have evolved alternate strategies to escape seed predators.

In the Lamto area, Don had found that 80 to 95 percent of trees and shrubs hid their seeds in edible fruits that served as bait to attract a variety of frugivorous bats, birds, monkeys, and rodents to carry seeds to new locations. This allowed seeds to escape virtually certain death beneath the parent tree, as well as permitting them to colonize new areas. Without such services, tropical plant communities worldwide would be very different.

Knowing which trees bats visited, it was easy for Don to hypothesize that bats had had a major role in structuring the Lamto plant community. Nevertheless, key questions remained before such a conclusion could be confirmed. What proportion of fruit from each plant species were bats eating? Did bats disperse seeds to appropriate locations? Were the seeds viable after passing through a bat's gut? And could the dispersed seeds escape seed predators long enough to germinate?

Don had designed a series of experiments. Initially, he focused on the cape fig, the first tree to colonize savannah areas. Over a period of several months, he had individually numbered 1,149 ripe fruits on 38 trees and checked them every day at dawn and dusk. Most figs disappeared within the first 24 hours, and 75 percent were removed at night, probably by bats.

Nevertheless, that did not necessarily mean that bats were responsible for the proliferation of cape figs into savannahs. In Don's next test, he put out hundreds of ten-foot-square plastic sheets along savannah transects. None were placed beneath trees, so all feces deposited were likely dropped by flying birds or bats. He predicted that if fruit bats were primary cape fig seed dispersers, most of the feces containing that species' seeds should fall at night.

In fact 94 to 100 percent of cape fig–containing feces were deposited at night. Though birds removed 25 percent of the fruit by day, they accounted for only 0 to 6 percent of seeds dropped in open areas. He found that birds spent more time feeding in fruiting trees, so fewer of their feces fell in openings. In the end Don documented that 95 percent of the "seed-rain" in open areas came from bats. In lab experiments he also demonstrated that fruit-eating bats, on average, consumed twice their body weight nightly, substantially more than most other frugivorous animals.

Subsequent research has shown that rainforest birds avoid open areas, preferring the protection of trees. However, at night, without fear of hawk attacks, fruit-eating bats readily cross openings. Also, in order to eat twice their body weight, these bats have extremely rapid digestion and save energy by defecating in flight. This makes them ideal dispersers of seeds into cleared areas throughout the world's tropics.

Many bat-dispersed plants are especially well adapted to surviving harsh, dry conditions and thus are referred to as "pioneer plants." Once established, such plants can provide food and shelter for birds who defecate mostly from perches, bringing seeds of less tolerant forest plants that need shade and moisture. Primates and other larger animals add final complexity. Only when all are present can a fully complex forest be established.

Don next compared germination success for hundreds of seeds removed from ripe fruits versus collected from bat feces. Half of those from bat feces sprouted in just six days, and all germinated within a few more. In contrast, no more than 10 percent of those taken directly from fruit ever germinated.

The story that emerged from Don's research is that fruit bats are extraordinarily effective reforesters of open areas on a continent where forest regrowth on abandoned farmland is often urgently needed.

Illustrative of just how important Africa's fruit bats can be, the colony of eight million straw-colored fruit bats living in Zambia's Kasanka National Park consumes approximately 6,000 tons of fruit and nectar from native trees and shrubs nightly, and seasonally migrates across enormous expanses of Africa. The full impact of such consumption is almost beyond comprehension.

United Kingdom bat expert Paul Racey noted that conserving straw-colored fruit bats "is crucial to the health of vast stretches of African forests and to the continent's timber industry." In Ghana, these bats account for more than 98 percent of seed dispersal for the iroko, West Africa's most valued timber tree. And in the Ivory Coast, Don Thomas estimated that, "In one night a single colony may disperse the seeds of nearly a half million pounds of iroko and other fruits throughout the surrounding forests."

In November 2003, I led a Bat Conservation International eco-tour to visit the spectacular Kasanka colony. Pregnant bats begin arriving in mid-October. Numbers peak in mid-November, and most depart for far-flung destinations by late December.

Led by an armed park guide, with bat sounds filling the air, our group entered the two-and-a-half-acre roosting area of evergreen swamp forest where the bats live. Unlike most Old World fruit bats, these form dense clusters. From a distance, they resemble great swarms of honeybees. We were warned not to step on dead branches, as cracking sounds would attract crocodiles from the nearby river.

There were so many bats crammed into such a small area that the sheer weight of them sometimes caused branches to break, plummeting whole clusters to the ground before they could fly. Listening crocodiles had learned to come on the run for a potentially easy meal. As I had seen at free-tailed bat caves in Texas, many predators were attracted to extra-large bat aggregations. In this case they included leopards, pythons, and black mamba snakes. It was the guide's duty to ensure that visitors did not disturb the mother bats too much as well as to minimize visitors' risks from predators.

We were escorted to ladders leading to blinds from which we could closely observe the bats.

Like many of their relatives, straw-colored fruit bats are quite handsome. Their eyes are large and hazel, faces foxlike, bodies straw-colored with a tan stripe down each side. And the fur of their necks ranges from yellowish to bright orange, especially in adult males.

The sight of so many animals, packed so densely together, was a once-in-a-lifetime experience, even for me. Each adult weighs only about three-quarters of a pound, but the combined weight of eight million exceeds six and a half million pounds. This is Africa's largest aggregation of mammals.

That evening we returned to witness the bats' exodus to feed. As the setting sun turned scattered clouds a bright orange, the sky dramatically filled with bats. Eight million flying mammals, each with a nearly three-foot wingspan, is an unforgettable sight.

The entire sky was filled for as far as we could see, 360 degrees around. Flapping slowly with long thumbs extended, they passed gracefully overhead, riveting our attention until it became too dark to see.

As more people come to appreciate the importance of such wonders of nature, I'm hopeful that sights like the Kasanka bat emergence will be protected far into the future. Nevertheless, I can't help but be concerned. When these bats leave Kasanka, they spread across countless thousands of square miles of Africa, facing a wide range of threats. They are often overhunted for bush meat.

At the Lamto Ecological Research Station reserve, Don Thomas once apprehended two poachers who had illegally shot a thousand of these bats in a single day's hunt. In local currency, those bats were worth the equivalent of $1,000 U.S., a small fortune for an average African worker.

To avoid shotgun-toting commercial hunters, straw-colored fruit bats learned long ago to live in city parks whenever possible. During peak occupancy, a park in the center of Abidjan was sheltering hundreds of thousands of these bats in relative safety. I say *relative*, because Don was quickly able to introduce me to teenage boy bat hunters relying on slingshots to kill bats, even in Abidjan. Driving slowly along a tree-lined street, bordered by high-rise apartment buildings, we quickly spotted a group of three boys in their early teens who were not very successfully attempting to blend in. As Don explained, each group included a marksman, a bagger, and a negotiator. The marksman had to be good with a slingshot. The bagger had to be an extra-fast runner, because shot bats didn't always fall quickly or in a location convenient for retrieval, exposing the waiting bagger to greater risk of apprehension. The negotiator was responsible for bribing guards whom homeowners paid to prevent bat hunting.

The high-rise buildings had large, easily broken windows, so occupants hired guards to keep slingshot hunters away. Nevertheless, the kids were clever. By offering the best bribes, they could entice guards to enter into lengthy chases of rival hunters so they could shoot bats unmolested for several minutes until they returned. The young hunters were happy to explain their strategy to Don and pose for my pictures. Don said the boys were likely making more money than their harder-working fathers did at their regular jobs.

After we had finished with the first interview and were driving slowly, trying to spot other bat hunters, we were stopped by a policeman. When Don explained that I was a bat researcher studying bat-hunting boys, the officer threatened to arrest me if I didn't promise never to tell anyone that people in Abidjan ate bats. Don, serving as my interpreter, just smiled and assured him I could keep the secret, knowing full well I'd have lots of fun sharing the story with friends.

Fun stories aside, the threats faced by straw-colored fruit bats and their flying fox relatives are serious and could jeopardize whole ecosystems and economies throughout most of Africa, Asia, Australia, and the Indian Ocean and Pacific Islands. All Old World fruit and blossom bats belong to a single family, the Pteropodidae, which includes approximately 200 species. Their common names are a bit confusing. Some are called fruit bats while others are referred to as blossom bats. The largest ones are simply lumped together as flying foxes. Most eat both fruit and nectar, however, and there is no agreement on how large a species must be to be called a flying fox. Since most feed at both fruit and blossoms and have foxlike faces, I find it easier to just think of the entire family as flying foxes.

These bats range in size from giants with six-foot wingspans to tiny ones that can fit in the palm of my hand. Some rank among the most colorful and strikingly marked mammals, and all are highly intelligent. In fact neurologist Jack Pettigrew discovered that their brains are remarkably similar to those of primates. In Australia, animal rehabilitators who have hand-reared orphaned babies report that the bats retain long-term memory of their benefactors, and that they periodically return from the wild with exuberant shows of affection.

Unfortunately, they also rank among our planet's most frequently persecuted and often endangered animals. Several island species became extinct because of overhunting for human food, long before I founded Bat Conservation International in 1982. Many populations have been dramatically reduced, especially in Australia where, as explained in the next chapter, they are still killed by fruit growers and additionally threatened by greatly exaggerated fear of disease. Recently, dire warnings of potential disease outbreaks from Africa's city-dwelling straw-colored fruit bats have put them in jeopardy despite the fact that countless thousands of Africans have shared their cities with these bats for hundreds of years without harm. People simply don't tolerate animals believed to be dangerous, regardless of the benefits the animals may provide.

As seen in the next two chapters, the threats faced by flying foxes, though serious, can be overcome by diplomatic education.

CHAPTER 14
FRUIT GROWER COMPLAINTS

"WHY ARE YOU LOOKING FOR BATS? Can you help us kill them?" A Kenyan farmer was responding to my inquiry regarding the possible existence of bat caves in the area.

"Why would you want to kill them?" I asked.

Looking a bit surprised by my response to what he considered to be a self-evident need, he said, "They eat our mangoes. They're a real problem!"

Having repeatedly heard similar responses, I decided to investigate. This man emphasized that he'd seen bats eating his fruit. "They even bite into green fruits!" When he picked one up, I showed him the tooth marks were from a monkey, not a bat.

To better understand the problem, I examined more than 7,000 mangoes during commercial harvests. Monkeys and their smaller relatives, the bush babies, accounted for the vast majority of problems being blamed on bats. The largest mango exporter in Kenya at the time, Akberkhan Khan, was the only farmer I met who understood the impact of primates versus bats in his orchards.

He had hired guards to chase away the monkeys. Regarding bats, he explained, "We have to pick and ship mangoes five to seven days before they are ripe. Bats are eating those that have ripened prematurely or have been missed by pickers. I'm happy to have them eat such mangoes because this limits breeding opportunities for fruit fly and fungus pests."

To test Akberkhan's belief, I caught several Egyptian fruit bats (*Rousettus aegyptiacus*) and three other species of small fruit bats as they came to feed on mangoes in his orchards. Then I released them in my portable, ten-foot-square studio. I repeatedly offered the bats freshly harvested, not-yet-ripe mangoes and other fruits, all of which the bats ignored even when they were extremely hungry. In contrast, they ravenously consumed ripe fruits.

Serious misunderstandings about fruit bats as orchard pests are widespread in the Old World tropics, but nowhere have erroneous conclusions caused greater devastation than in Australia. In 1929, the state governments of New South Wales and Queensland collaborated to hire a distinguished British biologist, Francis Ratcliffe, to "discover some wholesale method of destruction" which would once and for all eradicate flying foxes. He spent two years interviewing orchardists, thoroughly researching the problem, and he came to an astonishing conclusion: "The assumption that the flying fox is a menace to the commercial fruit industry of Australia is quite definitely false, and cannot be cited as a valid reason for expenditure of public money on control." He reported finding "a great deal of exaggeration," with orchardists often citing exceptional events as typical.

Ratcliffe estimated that the gray-headed flying fox (*Pteropus poliocephalus*) population had already declined by 50 percent since the arrival of European settlers. Yet, in his time, millions still remained. Unfortunately, his findings went unheeded and massive eradication campaigns followed. By 1990 bat biologists found fewer than 400,000 of these bats continent-wide.

More than 60 years after Ratcliffe's timely study, I had an opportunity to examine Australian problems. While featuring flying foxes for the *Secret World of Bats* documentary, Dieter Plage and I spent several days interviewing and filming Jim Trappel, a leading agricultural contractor and orchardist near Sydney.

Seeing he had covered most of his peach and nectarine orchards with netting, we stopped and introduced ourselves. When I asked if he had had trouble with flying foxes, he replied, "Those black devils have cost me plenty! They can wipe out eighty percent of my crop in just a few nights."

Over the next several days, we would become friends as I did my best to understand Jim's problems. He had been growing peaches and nectarines for 30 years. When I asked how often the bats caused problems, he admitted, "They mostly leave me alone. They prefer blossoms of gum trees. But when the blossoms fail in drought years, they can near destroy me."

When I asked how many times that had happened, Jim responded, "It's happened three times. The worst, we lost near 80 percent of our crop. Once it was 50 percent, and the other was nearly a third."

The next morning I noticed that colorful rosellas, relatives of parrots, were helping themselves to substantial numbers of fruits that weren't yet protected by netting. When I pointed them out, he commented, "Aren't they pretty?"

But I wondered, "From orchards you haven't yet netted, how much do you lose to them?"

When he estimated 10 percent a year, I couldn't resist quickly calculating that 10 percent a year added up to nearly twice the overall cost of bat losses. Jim grinned sheepishly and admitted he hadn't been completely objective. Then he went on to say, "You have to understand that we growers exaggerate a lot. When I see damage on one or two trees, I tell the wife we've been nearly wiped out. But in bad years, losses can near destroy small growers who live from crop to crop."

Jim was quick to point out that soon he'd have no bird or bat losses. In his first trials protecting his orchards with netting, he had avoided losses to both birds and bats, as well as to hail damage. Also, his crop had ripened earlier, enabling him to sell for higher prices. He had been so pleased he'd bought a distributorship, thinking all his neighbors would want to follow suit. But he lamented, "Most live from crop to crop and simply don't suffer often enough to convince them to invest."

Soon after returning home, I wrote to the Honorable Robert Carr, premier of New South Wales. Five years earlier, Dedee Woodside, curator of mammals at Sydney's Taronga Zoo, and I had met with him and convinced him to provide the first statewide protection for flying foxes. Based on what I'd learned from Jim Trappel, I suggested low-interest loans to farmers to protect orchards with netting, a political win-win since both farmers and conservationists should be pleased. I was delighted when such assistance was authorized.

But just as significant progress was being made, an Associated Press headline, dated October 21, 1996, announced the discovery of Hendra virus in Australia, proclaiming, "Fruit Bats Source of Fatal Virus." Educational progress of the previous decade was set back dramatically by this one speculative announcement. When I returned to Australia only a short time after the appearance of sensational headlines, people repeatedly warned me to stay away from flying foxes, because they were dangerous.

In subsequent decades, Hendra has proven to be one of the

continent's rarest sources of mortality, through early 2014 having accounted for an average of just four horse deaths annually and only two human deaths (contracted from sick horses) in total. Lab tests failed to transmit the virus from flying foxes to horses, and none of the hundreds of bat biologists and animal rehabilitators who have spent thousands of hours in close contact with flying foxes have contracted the disease. Only domestic housecats have proven capable of transmitting Hendra to horses.

Nevertheless, despite the development of a new vaccine and the Australian Veterinary Association's warnings that killing flying foxes would be counterproductive, the New South Wales government announced a new policy allowing city councils to kill bat colonies, beginning in October 2014.

Convincing people to be concerned about flying fox decline is challenging when thousands can still be seen in single roosts. They often appear to be common even when they remain only as tiny remnants of former populations, remnants that are no longer sufficient to serve forest needs for seed dispersal and pollination. Thousands of seeds may have to be carried to new locations to result in one that survives to become a new tree.

Despite the sometimes dire circumstances fruit bat populations face, there are examples of progress. Prior to my study in Kenya, the Israeli government had treated all of its known bat caves with a persistent organochlorine pesticide known as lindane, hoping to eradicate Egyptian fruit bats. They inadvertently exterminated whole populations of insect-eating bats; an explosion of moth pests followed. When Israeli biologists shared my research with government officials and mango farmers, more than a decade of poisoning was ended.

In my experience, most farmers and government officials are

open to sound advice on crop protection. When we take time to adequately investigate complaints and suggest possible remedies, amazing progress often results. In American Samoa I would find that, once befriended and listened to, even commercial hunters could be successfully recruited as allies in flying fox conservation.

CHAPTER 15

A NATIONAL PARK FOR BATS

IT WAS MY FIRST EVENING on the American Samoa island of Tutuila, and my hunter companions were disappointed to have shot just two bats. "It's too bad you couldn't have been here a year ago," they said. "In only an hour, three of us could get a hundred or more." They went on to enthusiastically explain their techniques and experience. The bats made easy targets as they flew low through passes in 1,000-foot-tall ridges that bisect the island. And I soon learned that the hunters were successful businessmen who simply sold bats commercially to pay for their shotgun shells.

As we were returning to their vehicles, we were accosted by two American Peace Corps volunteers who berated us severely for having virtually wiped out the island's flying foxes. Though I carried no gun, they singled me out as a mainlander who should have known better. Although I sympathized with their concerns, I also am not opposed to hunting. In fact, in a modern world, game animals often receive the best protection, thanks to the concerns of responsible hunters. I don't like to see flying foxes shot, but I also don't like thinking of them going extinct.

The Peace Corps volunteers weren't the only ones who were upset that night. Paul Cox, a leading Pacific Island botanist, was surprised and outraged. He had asked me to come here to help save Samoan flying foxes (*Pteropus samoensis*). The Chamorro people of Guam had eaten the Guam flying fox (*Pteropus tokudae*) into extinction and had forced the Mariana flying fox (*Pteropus mariannus*) into endangered status before turning to commercial importation from other islands to satisfy their appetites. Now the trade had reached American Samoa, where Paul had seen an alarming decline and feared extinction.

Paul had raised a Samoan flying fox that had been orphaned by hunters, and he knew just how cute, charming, and affectionate they could be. The last thing he wanted was to see them being killed. Nevertheless, when I explained that the hunters I'd met had admitted being concerned about the dramatic decline of flying foxes and might even be persuaded to become allies in our mission to save them, he finally agreed to help. And no one could be more helpful when he wanted to be.

The next morning, Paul shared one of his favorite places with me. Seated atop Alava Ridge, approximately 1,600 feet above the Pacific Ocean on Tutuila Island, I was mesmerized, not only by the spectacular view, but also by my first encounter with a Samoan flying fox, a handsome species with a three-foot wingspan, dark brown body, contrasting bright golden-orange neck and shoulders, and a head crowned in silvery white. Riding on thermal breezes, like a large bird, it repeatedly soared past, sometimes within just a few feet. Samoa is one of the few places on earth where one can observe such giants soaring alongside dainty fairy terns in midmorning sun.

Paul and I were seated on a 15-foot-wide grassy perch atop a rocky pinnacle. Sheer cliffs dropped off on both sides, and we could see for miles in all directions. Pristine cloud forest gave way to rainforest far below, showcasing a wide variety of lianas, orchids, tree ferns, and other tropical plants, some 30 percent of which represented species found nowhere else on earth. It was easy to see how Paul had fallen in love with this special place and its flying fox guardians.

Little did we realize that someday we'd take great pride in knowing that these bats and this place had become part of a national park that former commercial hunters had helped us create.

Having a wonderful way with people, understanding the local culture, and fluently speaking the chief's dialect of Samoan, Paul was a formidable ally. The next evening, he did a masterful job of hiding his feelings as he watched the hunters kill half a dozen more of these beautiful bats. And with his help, by the evening's end, I was able to explain how hunters often had become leading conservationists in the mainland United States, ensuring good hunting not only for themselves but for future generations.

By the second night, the hunters had enthusiastically agreed to have us represent their interests when we met with Governor Lutali. In just seven months, legislation was drafted and passed that included hunting seasons and bag limits and prohibited further commercial hunting, which the hunters themselves viewed as a major part of the problem.

Friends who knew Samoans better than I did had warned that the hunting legislation would just be ignored, but having had a voice in its passage, most hunters cooperated fully, and commercial hunting did end. The hunters even voluntarily declared a five-year moratorium on flying fox hunting, so that the bats could recover. And when a typhoon stripped trees of fruit and flowers that the bats needed to survive, villagers set up flying fox feeding stations and extended the moratorium.

Some people may be shocked that I would support flying fox hunting for sport, but the alternative would have meant almost certain failure to save them. Through our collaboration with hunters, we ended up gaining far more than we ever could have anticipated.

Some 80 percent of Samoa's lowland rainforests had been lost since the beginning of human settlement 3,000 years earlier, and it soon became obvious that the survival of Samoa's flying foxes would require more than mere protection from commercial hunting. Like all animals, they needed habitat, and that was disappearing rapidly. When we shared our concerns with the hunters, they strongly agreed.

However, land in Samoa is communally owned and cannot be sold, not even to protect habitat. The idea of a national park was discussed, but it seemed like an impossible dream. Nevertheless, Verne and Marion Read funded several trips through BCI for Paul Cox to discuss possibilities with our hunter friends and local leaders. And Verne, Paul, and I also met with members of the United States Congress to suggest the idea.

By late in 1986 we'd gained support from Samoan chiefs, as well as from the United States National Park Service, sufficient to warrant congressional hearings. We were informed, however, that Congress simply didn't have the funds required to fly congressmen to Samoa for hearings. Verne, not to be denied, responded that if that was the only problem, he'd personally fund the hearings. Fortunately, that seemed to tip the balance, and he and Marion didn't actually have to pick up what undoubtedly would have been a very substantial tab.

The proposed park would include parts of three separate islands, and hearings were held on all of them. On each one we had to sit cross-legged on pandanus mats through seemingly endless kava ceremonies, as all the chiefs and other dignitaries spoke in Samoan, with Paul Cox translating. A kava cup was passed around according to strictly orchestrated tradition. The kava was quite a potent substance. Drunk like tea, it deadens the mouth like Novocain. It is definitely an acquired taste. The food at the feasts was delicious and plentiful, but I found it exceedingly difficult to sit cross-legged with my feet tucked beneath me for long periods. The ever-thoughtful Samoans finally realized that some of us were in extreme discomfort and brought additional pandanus mats so we could sit with our feet forward without a breach of etiquette.

The more enjoyable parts of the hearings were our tours of gorgeous areas proposed as parklands, from pristine, white-sand beaches to spectacular mountain ridges. However, some of the planes we flew on seemed to be in an alarming state of disrepair. I'll never forget the flight to Tau. I sat next to a young congressional aide who apparently had never been in a small plane before. As she looked up to see insulation and loose electrical wiring hanging from the fuselage, I couldn't resist teasing her by pointing out that at least they had used stainless steel baling wire to hold the engine covers in place!

Based on a favorable response in Samoa, and the findings of visiting congressmen, a feasibility study for the park was authorized, and congressional hearings were held in Washington, D.C. Paul even brought his hand-reared Samoan flying fox, and the bat was a big hit.

Two Samoan high chiefs, dressed in traditional regalia, were also a hit. Paul Cox served as their translator, with sometimes amusing challenges, such as when one of the chiefs, in an effort to illustrate the importance of the hearing for Samoans, stated that it was "as sacred as the mating of sea turtles." Paul, understanding how difficult it might be for Americans unacquainted with Samoan cultural traditions to keep a straight face, opted to translate simply that they considered this meeting to be of very special importance.

Several congressional staffers considered the hearing to be so successful they called it a "love-in." However, by the time the park bill gained unanimous House approval, very little time remained for getting it introduced and passed by the Senate before their October recess. BCI had only a few thousand members at the time, but they included people of exceptional influence who were willing to work diligently with staff. For the next month we virtually lived on the phone, enlisting the support of senators from coast to coast in a strong bipartisan effort.

In order to obtain Senate approval before Congress adjourned, our Washington collaborators got the bill included in an omnibus package. With only two weeks to go, however, the omnibus package failed.

To our amazement, the Senate, in the nick of time, passed our bill separately and unanimously on October 12. We had apparently been far more influential than any of us dreamed possible. We were elated! What could go wrong now? Our bill had passed unanimously in both the House and the Senate.

Just as we were about to celebrate, I received a call from the Republican advisor on Pacific Island Affairs, Manase Mansur, who had played a key role in the final weeks of lobbying. He said he had some very bad news. President Reagan had decided to pocket-veto the park bill (simply not sign it), and there was nothing anyone could do about it.

I was stunned. I suggested that, because we were in an election year, there should be significant political capital for the Republican Party to have played a leadership role in creating the first national park in American history to protect a tropical rainforest.

Manase agreed that I had a point and directed me to an insider in the Bush campaign. It turned out that Congressman Robert Lagomarsino, who had cosponsored the bill, was in an extremely tight race in California in a district where people were quite concerned about environmental issues. Moreover, the Bush campaign feared that if Lagomarsino lost, Republicans could lose the state.

To make a long story short, though I am strictly nonpartisan, I ended up writing media pieces for the Lagomarsino campaign, extolling the congressman's leadership role in cosponsoring a bill to create America's first national park to protect a tropical rainforest. And apparently the Bush campaign did have some influence, because with no time to spare, President Reagan signed the park bill on October 31, 1988. Congressman Lagomarsino won reelection by one of the narrowest margins in history.

Legislation, though, was only the first step in creating a park. Four years after the National Park of American Samoa was approved by Congress and the president, still there was no park in Samoa. Several villages on three islands had to sign long-term leases. And although the park concept remained popular, the National Park Service's negotiations with traditional high chiefs had bogged down. A major stumbling block was that virtually no one in Samoa had actually experienced a national park. Some people even thought it was a place for sports events. In order to explain to their constituents, the high chiefs needed to experience a national park for themselves.

By 1992, it was clear to all concerned that our dream of a national park would be in serious trouble without prompt action. The National Park Service proposed bringing appropriate chiefs to Hawaii to experience a national park and how it could protect both natural and cultural heritage.

Once again, Verne and Marion Read came to the rescue. With

a grant to BCI from their family's Chapman Foundation, they funded 21 Samoan high chiefs, a princess, and government officials, the most influential delegation ever to leave Samoa, on a four-day tour of Hawaiian national parks.

Our Hawaiian hosts gave them a royal welcome, with traditional ceremonies, feasts, and tours of national parks and monuments on the islands of Hawaii and Oahu. The chiefs were able to see firsthand how a park could protect wildlife and preserve cultural heritage in Samoa. Hosting these chiefs was a rare privilege. I will always regret not having been able to record the thousands-of-years-old traditional songs they serenaded us with as we bused them among destinations. The princess in the group was quite amused, explaining that some of the songs were far too ribald to be sung at home. Whatever the content, we were mesmerized by what sounded to us like the best old-fashioned "barbershop quartet" singing we'd ever heard.

With understanding and enthusiasm, the high chiefs returned to Samoa and, encouraged by their hunter constituents, made the National Park of American Samoa a reality.

In 2013, while attending the 16th International Bat Research Conference in Costa Rica, I met Adam Miles, a bat biologist from the Department of Marine and Wildlife Resources in American Samoa, and was delighted to hear that Samoa's flying foxes have recovered.

I know that Paul joins me in hoping that future generations will find our special spot atop the Alava Ridge and enjoy their first encounter with a giant day-flying bat as much as we did.

EPILOGUE

Hope for the Future

"WE NEED HELP. We have bats!"

"Are they causing a problem?"

"My wife saw a bat in our backyard last night."

"What was it doing?"

"It was swooping around our children. Thank God she was able to get them inside before they were attacked!"

"What makes you think a bat would attack?"

"We read all about them in a women's magazine. It warned that many are rabid and that they especially like to attack children. It also said that if they get into your attic they breed like rabbits and can be very difficult to get rid of. We called a pest control company, and they offered to treat our attic with chemicals to keep them out. But they charge $500 and we can't afford it. We called the museum, and the receptionist forwarded our call. She said you're an expert."

Such calls to my office at the Milwaukee Public Museum were typical in the 1970s and early 1980s. In our city alone, individual pest control companies were earning up to $500,000 per summer placing chlorophacinone tracking powder or methyl bromide in area homes to "protect" families from bats. Though the chemicals were extremely hazardous to people, the business was lucrative.

As the museum mammalogist I got all the bat calls. I'd point out that I had spent a lifetime studying bats, often surrounded by thousands at a time, and had never been attacked, nor had I encountered a single instance in which a reported attack had proven true.

One family's horror story was prominently featured in local papers, generating seemingly endless panicked calls to my office. They claimed that the recent federal ban on the use of DDT had led to a bat population explosion and that their kids could no longer play in their lakeshore yard in the evening, nor sleep at night, for fear of bat attacks. "Their nightmares were terrible!"

Fed up with seeing so much misinformation published, I finally looked up their address and went to see them on a weekend evening. Unannounced, I knocked on their door. When an attractive, middle-aged woman responded, with a stern "Go away. I'm not buying" look in her eye, I realized I'd arrived at a bad time. She and her husband were hurriedly preparing for a dinner party.

I apologized for the interruption, explaining that I was a biologist from the museum, that I studied bats, had read about their predicament in the paper, and was wondering if I could help. Their rush to get rid of me was immediately forgotten as they poured out their story, assuming that I, of all people, would be sympathetic. And I was. From just a few responses to my questions, I could see they had been victimized by well-intended but naive media warnings, not by the bats.

When I had convinced them, the mother sincerely apologized for having been so wrong, but asked, "What should we tell our children? They're terrified." Just then a mischievous-looking, freckle-faced boy of about ten and his younger sister showed up, curious to see what was happening.

Unbeknownst to the parents, I'd been holding a little brown myotis (*Myotis lucifugus*) concealed in my left hand the whole time we'd been talking. As I slowly raised my hand and asked, "Would you like to meet a real bat?" there were audible gasps. But when the now very curious children leaned forward for a better look, the daughter exclaimed, "Oh, it's cute!" and the boy wanted to know if he could pet it.

After allowing them to overcome any lingering fear, I warned them that they should never attempt to handle bats on their own, because those they could catch would be far more likely than others to be sick and might bite in self-defense. But I also emphasized that even sick bats rarely bite people if left alone and that *any* animal bite, even from a dog or cat, could be a potential exposure to rabies that should be reported to their family doctor.

Like most families I spoke with in those days, they had read scary bat-attack stories. I explained that most publishers simply didn't know that such stories were false or grossly exaggerated. They just knew that sensational headlines attracted readers. Those who submitted such stories found them exceedingly helpful in promoting pest control business and public health budgets, despite the fact that such funds were being misdirected away from far more serious health threats. I pointed out that this would stop only when more readers became aware and complained to publishers.

The family offered to help and were quite disappointed when I couldn't stay and show off my bat as entertainment for their party.

I learned many good lessons from my early experiences combating misinformation. First, there are good people in all professions. I was able to recruit honest pest control operators to whom I referred customers with legitimate nuisance problems, and they gained even more business by helping me provide daylong workshops for public health personnel. They learned that where bat colonies were an actual nuisance, simple exclusion was by far the safest, longest-lasting remedy.

When I founded Bat Conservation International, honest members from among both pest control operators and public health personnel joined me in supporting statewide consumer protection legislation outlawing the poisoning of bats. And that success led to a national ban. It's amazing how much can be accomplished when we convert potential enemies into allies.

Many challenges remain, however. Bat populations from Asia and Africa to the Pacific Islands and Madagascar are still at risk from overhunting for bush meat. Many of Australia's flying foxes are severely threatened by a combination of human intolerance and climate change. Careless operation of wind energy facilities poses an extreme threat throughout the industrialized world, with nearly one million bats already being killed annually in North America alone. In addition, millions of American bats that hibernate in caves have been killed by a recently introduced fungal disease known as white-nose syndrome (WNS).

Added to these well-documented threats, we are seeing a big resurgence of exaggerated claims that bats are especially dangerous sources of so-called emerging diseases, such as Ebola, Marburg, Hendra, Nipah, and SARS. These diseases are not new – they're just so rare that they are only now being discovered. Their combined impact is minuscule compared to almost any other source of human mortality. Those who profit from speculation about potential bat-triggered pandemics can't explain the fact that hundreds of bat researchers worldwide remain healthy despite lifetimes of close association with bats. We are not protected against any of these diseases, although, like veterinarians, we are vaccinated against rabies in case of an occasional bite during handling. If these so-called emerging diseases are as dangerous as claimed, it's amazing that none of us has contracted one. In fact, even among the millions of people who eat bats or harvest guano from caves, reports of possible harm are exceedingly rare.

Despite all of these threats, I am increasingly optimistic. Following centuries in which bats have been objects of fear and disgust because of ignorance and resulting superstitions, a new generation of young scientists is finally shedding much-needed light on bats as essential allies and safe neighbors. In recent decades, hundreds of new bat species and billions of dollars' worth of bat contributions to human economies have been discovered.

Modern bat biologists can track bat movements with tiny radio transmitters, record their behavior on inexpensive low-light video cameras, identify them remotely by their echolocation calls, and even document the bugs they eat based on genetic typing of tiny fecal fragments. Long neglected, bat research is now becoming one of biology's most exciting frontiers.

In the long run, scientific advances, combined with growing opportunities for the public to experience bats close-up, bode well for bats. Studies of human attitudes reveal that the more we know about bats, the more we like them. And like them or not, new discoveries increasingly justify bat conservation based solely on economics. As people in Austin, Texas, have learned, even simple bat watching can be very good business. In recent decades, bat-watching opportunities have proliferated worldwide, and growing numbers of communities and their bat-watching visitors are discovering firsthand that bats make good neighbors.

We now live in an age of mass communication, which presents growing opportunities to spread the truth about bats. Even the blight from WNS has a silver lining. It has generated extraordinary publicity and awareness of the ecological and economic values of bats and has gained support for the first long-term monitoring of bat status trends in America, an important step toward improved protection.

WNS kills cave-hibernating bats by forcing them to exhaust stored fat reserves before spring. We are unlikely to find a cure or slow its spread, and the consequences are indeed tragic, but even among the most severely impacted species, some appear to be surviving with immunity and are beginning to gradually rebuild populations. WNS has in recent years invaded gray myotis, but to date there is no evidence of a major die-off.

My greatest concern is that, in our efforts to help, we don't inadvertently make matters worse. This is a time when already-stressed hibernating bats must be strictly protected from human disturbance. It is understandable that people want to help, and we can. For far too long, most efforts to protect and restore America's bat hibernation caves have been limited to those occupied by endangered species. This is an ideal time to extend help to important hibernation sites for all species and to initiate long overdue status monitoring across the landscape.

Wind energy production also poses a significant threat. But even here there is hope for an improved future. In 2003, we formed the Bats and Wind Energy Cooperative to address unexpectedly high bat kills at wind turbines. Managed by Bat Conservation International in collaboration with the United States Fish and Wildlife Service, the American Wind Energy Association, and the National Renewable Energy Laboratory (United States Department of Energy), it has become a model for collaboration among conservationists, scientists, and industry.

Scientists from around the world have partnered with corporate managers in search of solutions. The first scientifically credible methodology for preconstruction assessment and post-construction monitoring at wind energy sites has been developed, and substantial progress has been made in finding cost-effective solutions to reduce bat kills. Collaborating scientists have discovered that most bat mortality occurs on low-wind nights when energy production is insubstantial, and that it is largely limited to relatively short periods when bats migrate.

Just by reprogramming computers to activate turbines at slightly higher wind speeds during brief periods of increased risk, bat mortality could be greatly reduced at a cost of less than 1 percent of total energy production. And ongoing research is refining knowledge needed to better predict high-risk periods or repel bats. A relatively few pioneering companies have borne most of the cost of sharing standardized information with scientists, as well as testing new procedures for minimizing harm to bats, and these are greatly appreciated.

The next important step is to develop a means whereby concerned investors can be encouraged to preferentially support companies with the best track records. No company will initially have a perfect record of environmental responsibility, but if one is even 10 percent better than another, preferential investment can lead to greater gains. Little by little, such incentives can move mountains, possibly even helping solve broader issues like global climate change.

Early in my career, the plight of bats seemed hopeless. Throughout large areas of the world, especially in the United States and Canada, bats were generally viewed as dangerous vermin. Like the people who called me in Milwaukee, virtually everyone "knew" that bats were mostly rabid and would attack people. Even the world's leading conservation organizations avoided bats like the plague, believing them to be far too unpopular to be helped.

Bats were going extinct without even being recognized as endangered, and America's leading experts were predicting that the gray myotis, the focus of my early studies, would soon itself become extinct.

As I learned more about bats and their critical roles, I couldn't resist speaking out on their behalf. I began with owners of caves where I conducted my research. They were often amazed to learn that bats had a far better record of living safely with humans than even our beloved dogs, and that they also play essential roles in supporting human economies.

Most bat haters quickly changed their minds if I simply listened to their fears and responded with helpful facts from my own experience. On balance, bats are such powerful contributors to human well-being that converting people to like them can be amazingly simple.

In founding an organization on behalf of some of the world's most feared animals, one of the first lessons I had to learn was how to convert enemies from the past into allies for the future. Within virtually every group, from commercial hunters to energy-producing corporations, I have found good individuals who have taken pride in helping expand beachheads of progress.

As repeatedly illustrated in previous chapters, we've together made enormous progress, not just for bats, but for people too, and it has been done mostly by just winning friends. Together, we have sponsored hundreds of research projects to document bat values and needs, dramatically improved public perceptions of bats, and gained protection for thousands of critical bat roosts and other habitats, including the most important remaining nursery and hibernation caves in North America and a national park in American Samoa.

Most important, these actions have led to well-documented population recovery at numerous locations from Tennessee to Thailand. We now have millions more endangered gray myotis than when their extinction was predicted in the late 1960s and early 1970s.

I'd like to think that the success of our positive approaches will inspire a new generation to make even greater progress. Certainly, if so few of us could accomplish so much in such a short period of time, with so few resources, on behalf of such a seemingly hopeless cause, there must be hope for the future.

BIBLIOGRAPHY

INTRODUCTORY READING

Adams, R. A. *Bats of the Rocky Mountain West: Natural History, Ecology and Conservation.* Boulder, CO: University Press of Colorado, 2003.

Ammerman, L. K., C. L. Hice, and D. J. Schmidly. *Bats of Texas.* College Station, TX: Texas A&M University Press, 2012.

Fenton, M. B., and N. B. Simmons. *Bats: A World of Science and Mystery.* Chicago and London: University of Chicago Press, 2014.

Harvey, M. J., J. S. Altenbach, and T. L. Best. *Bats of the United States and Canada.* Baltimore, MD: Johns Hopkins University Press, 2011.

Schmidt-French, B. A., and C. A. Butler. *Do Bats Drink Blood?* New Brunswick, NJ: Rutgers University Press, 2009.

Tuttle, M. D. *America's Neighborhood Bats.* Austin, TX: University of Texas Press, 1988 (rev. ed. 2005).

CHAPTER 1: TEENAGE DISCOVERIES

Locke, R. "The Gray Bat's Survival." *BATS* 20, no. 2 (2002): 4–9.

Pringle, L. *Batman: Exploring the World of Bats.* New York: Charles Scribner's Sons, 1991.

Tuttle, M. D., and D. E. Stevenson. "Variation in the Cave Environment and Its Biological Implications." In *1977 National Cave Management Symposium Proceedings,* edited by R. Zuber, J. Chester, S. Gilbert, and D. Rhoads. Albuquerque, NM: Adobe Press, 1978, 108–21.

CHAPTER 2: SAVING THE GRAY MYOTIS

Barbour, R., and W. H. Davis. *Bats of America*. Lexington: University of Kentucky Press, 1969.

Kerth, G., N. Perony, and F. Schweitzer. "Bats Are Able to Maintain Long-Term Social Relationships Despite the High Fission-Fusion Dynamics of Their Groups." *Proceedings of the Royal Society B* 278, no. 1719 (Sept. 2, 2011): 2761–67. doi:10.1098/rspb.2010.2718.

Locke, R. "The Gray Bat's Survival." *BATS* 20, no. 2 (2002): 4–9.

Pinkley, J. E. *Fern Cave: The Discovery, Exploration and History of Alabama's Greatest Cave*. Taft, TN: Blue Bat Books, 2014.

Stevenson, D. E., and M. D. Tuttle. "Survivorship in the Endangered Gray Bat (*Myotis grisescens*)." *Journal of Mammalogy* 62 (1981): 244–57.

Tuttle, M. D. "Population Ecology of the Gray Bat (*Myotis grisescens*): Factors Influencing Preflight Growth and Development." *Occasional Papers of the Museum of Natural History, University of Kansas* 36 (1975): 1–24.

———. "Population Ecology of the Gray Bat (*Myotis grisescens*): Factors Influencing Growth and Survival of Newly Volant Young." *Ecology* 57 (1976): 587–95.

———. "Population Ecology of the Gray Bat (*Myotis grisescens*): Philopatry, Timing and Patterns of Movement, Weight Loss During Migration, and Seasonal Adaptive Strategies." *Occasional Papers of the Museum of Natural History, University of Kansas* 54 (1976): 1–38.

———. "Status, Causes of Decline, and Management of Endangered Gray Bat." *Journal of Wildlife Management* 44 (1979): 955–60.

Tuttle, M. D., and D. E. Stevenson. "An Analysis of Movement as a Mortality Factor in the Gray Bat, Based on Public Recoveries of Banded Bats." *American Midland Naturalist* 97 (1977): 235–40.

CHAPTER 3: TRACKING BAT NIGHTLIFE

Brady, J., T. H. Kunz, M. D. Tuttle, and D. Wilson. *Gray Bat Recovery Plan*. Denver, CO: U.S. Fish & Wildlife Service, 1982.

Rabinowitz, A. R. "Habitat Use and Prey Selection by the Endangered Gray Bat, *Myotis grisescens*, in East Tennessee." (master's thesis, University of Tennessee, Knoxville, 1978).

CHAPTER 4: INVESTIGATING VAMPIRE BATS

Belwood, J. J., and P. A. Morton. "Vampires: The Real Story." *BATS* 9, no. 1 (1991): 11–16.

Brown, D. E. *Vampiro: The Vampire Bat in Fact and Fantasy.* Silver City, NM: High-Lonesome Books, 1994.

Czaplewski, N. J., and W. D. Peachy. "Late Pleistocene Bats from Arkenstone Cave, Arizona." *Southwestern Naturalist* 48, no. 4 (2003): 597–609.

Grayson, D. K. "Late Pleistocene Mammalian Extinctions in North America: Taxonomy, Chronology, and Explanations." *Journal of World Prehistory* 5, no. 3 (1991): 193–231.

Greenhall, A. M., and U. Schmidt (eds.). *Natural History of Vampire Bats.* Boca Raton, FL: CRC Press, 1988.

Gut, J. "A Pleistocene Vampire Bat from Florida." *Journal of Mammalogy* 40, no. 4 (1959): 534–38.

Martin, R. A. "Synopsis of Late Pliocene and Pleistocene Bats of North America and the Antilles." *American Midland Naturalist* 87, no. 2 (192): 326–35.

McNab, B. K. "Energetics and the Distribution of Vampires." *Journal of Mammalogy* 54, no. 1 (1973): 131–44.

Medcalf, R. L. "Desmoteplase: Discovery, Insights and Opportunities for Ischaemic Stroke." *British Journal of Pharmacology* 165 (2011). doi:10.1111/j.1476-5381.2011.01514.x. http://www.ncbi.nlm.nih.gov/pubmed/21627637.

Olsen, S. J. "Additional Remains of Florida's Pleistocene Vampire." *Journal of Mammalogy* 41 (1960): 458–62.

Peters, S. "Banishing the Vampires of the Jungle." *BATS* 22, no. 2 (2004): 1–3.

Schutt, B. *Dark Banquet: Blood and the Curious Lives of Blood-Feeding Creatures.* New York: Crown, 2008.

Turner, D. C. *The Vampire Bat: A Field Study in Behavior and Ecology.* Baltimore, MD: Johns Hopkins University Press, 1975.

Winter, M., and C. Coen. "Lure of the Vampires." *BATS* 15, no. 2 (1997): 7–10.

CHAPTER 5: BATS THROUGH A CAMERA'S EYE

Allen, T. B. (ed). *Wild Animals of North America.* Washington, DC: National Geographic Society, 1979.

Hensley, D. "Adventures in Photographing Bats." *BATS* 10, no. 3 (1992): 5–9.

Tuttle, M. D. "Gentle Fliers of the African Night." *National Geographic* 169, no. 4 (1986): 540–558.

———. "Saving North America's Beleaguered Bats." *National Geographic* 188, no. 2 (1995): 37–57.

CHAPTER 6: DISCOVERING FROG-EATING BATS

Barclay, R. M., M. B. Fenton, M. D. Tuttle, and M. J. Ryan. "Echolocation Calls Produced by *Trachops cirrhosus* (Chiroptera: Phyllostomidae) While Hunting for Frogs." *Canadian Journal of Zoology* 59 (2009): 750–53.

Dapper, A. L., A. T. Baugh, and M. J. Ryan. "The Sounds of Silence as an Alarm Cue in Túngara Frogs, *Physalaemus pustulosus*." *Biotropica* 43, no. 3 (2011): 380–85.

Jones, P. L., M. J. Ryan, V. Flores, and R. A. Page. "When to Approach Novel Prey Cues? Social Learning Strategies in Frog-Eating Bats." *Proceedings of the Royal Society B* 280 (October 2013): 20132330. http://dx.doi.org/10.1098/rspb.2013.2330.

Page, R. A., and M. J. Ryan. "Flexibility in Assessment of Prey Cues: Frog-Eating Bats and Frog Calls." *Proceedings of the Royal Society B* 272 (April 2005): 841–47.

———. "Social Transmission of Novel Foraging Behavior in Bats: Frog Calls and Their Referents." *Current Biology* 16 (2006): 1201–05.

Ryan, M. J., and M. D. Tuttle. "Bat Predation and Sexual Advertisement in a Neotropical Anuran." *American Naturalist* 119 (1982): 136–39.

———. "The Ability of the Frog-Eating Bat to Discriminate Among Novel and Potentially Poisonous Species Using Acoustic Cues." *Animal Behaviour* 31 (1983): 827–33.

Ryan, M. J., M. D. Tuttle, and R.M.R. Barclay. "Behavioral Response of the Frog-Eating Bat, *Trachops cirrhosus*, to Sonic Frequencies." *Journal of Comparative Physiology* 150 (1983): 413–18.

Ryan, M. J., M. D. Tuttle, and L. K. Taft. "The Costs and Benefits of Frog Chorusing Behavior." *Behavioral Ecology and Sociobiology* 8 (1981): 273–78.

Tuttle, M. D. "The Amazing Frog-Eating Bat." *National Geographic* 161, no. 1 (1982): 78–91.

Tuttle, M. D., and M. J. Ryan. "Bat Predation and the Evolution of Frog Vocalizations in the Neotropics." *Science* 214, no. 4521 (1981): 677–678.

———. "The Role of Synchronized Calling, Ambient Light, and Noise in Anti-Bat-Predator Behavior of a Tree Frog." *Behavioral Ecology and Sociobiology* 11 (1982): 125–131.

Tuttle, M. D., M. J. Ryan, and J. J. Belwood. "Acoustical Resource Partitioning by Two Species of Phyllostomid Bats (*Trachops cirrhosus* and *Tonatia sylvicola*)." *Animal Behaviour* 33 (1985): 1369–71.

Tuttle, M. D., L. K. Taft, and M. J. Ryan. "Evasive Behavior of a Frog in Response to Bat Predation." *Animal Behaviour* 30 (1982): 393–97.

CHAPTER 7: FINDING AMERICA'S MOST ELUSIVE BATS

Belwood, J. J., and J. H. Fullard. "Echolocation and Foraging Behavior in the Hawaiian Hoary Bat, *Lasiurus cinereus semotus*." *Canadian Journal of Zoology* 62 (1984): 2113–20.

Chebes, L. "*Eumops perotis* (Western Bonneted Bat)." Animal Diversity Web, University of Michigan Museum of Zoology, 2002. http://animaldiversity.ummz.umich.edu/accounts/Eumops_perotis/.

Fenton, M. B. "Sound Wars: How Bats and Bugs Evolve New Weapons and Defenses." *BATS* 20, no. 1 (2002): 1–4.

Filla, J. "*Myotis keenii* (Keen's Myotis)." Animal Diversity Web, University of Michigan Museum of Zoology, 2005. http://animaldiversity.ummz.umich.edu/accounts/Myotis_keenii/.

Fullard, J. H., and J. W. Dawson. "The Echolocation Calls of the Spotted Bat *Euderma maculatum* Are Relatively Inaudible to Moths." *Journal of Experimental Biology* 200 (1997): 129–37.

Hussain, S. "*Euderma maculatum* (Spotted Bat)." Animal Diversity Web, University of Michigan Museum of Zoology, 2002. http://animaldiversity.ummz.umich.edu/accounts/Euderma_maculatum/.

Leonard, M. L., and M. B. Fenton. "Habitat Use by Spotted Bats (*Euderma maculatum, Chiroptera: Vespertilionidae*): Roosting and Foraging Behaviour." *Canadian Journal of Zoology* 61 (1983): 1487–91.

———. "Echolocation Calls of *Euderma maculatum (Vespertilionidae)*: Use in Orientation and Communication." *Journal of Mammalogy* 65 (1984): 122–26.

Woodsworth, G. C., G. P. Bell, and M. B. Fenton. "Observations of the Echolocation, Feeding Behavior, and Habitat Use of *Euderma maculatum (Chiroptera: Vespertilionidae)* in Southcentral British Columbia." *Canadian Journal of Zoology* 59 (1981): 1099–1102.

CHAPTER 8: CACTI THAT COMPETE FOR BATS

Fleming, T. H., S. Maurice, S. L. Buchmann, and M. D. Tuttle. "Reproductive Biology and Relative Male and Female Fitness in a Trioecious Cactus, *Pachycereus pringlei (Cactaceae)*." *American Journal of Botany* 81, no. 7 (1994): 858–67.

Fleming, T. H., M. D. Tuttle, and M. A. Horner. "Pollination Biology and the Relative Importance of Nocturnal and Diurnal Pollinators in Three Species of Sonoran Desert Columnar Cacti." *Southwestern Naturalist* 41, no. 3 (1996): 257–69.

Horner, M. A., T. H. Fleming, and C. T. Sahaley. "Foraging Behavior and Energetics of a Nectar-Feeding Bat, *Leptonycteris curasoae* (Chiroptera: Phyllostomidae)." *Journal of Zoology London* 244 (1998): 575–86.

Sahaley, C. T. "Peru's Bat-Cactus Connection." *BATS* 13, no. 3 (1995): 6–11.

Tuttle, M. D. "Bats, the Cactus Connection." *National Geographic* 179, no. 6 (1991): 130–40.

CHAPTER 9: FREE-TAILED BAT CAVES AND CROP PESTS

Balcombe, J. P. "Vocal Recognition of Pups by Mother Mexican Free-Tailed Bats, *Tadarida brasiliensis Mexicana*." *Animal Behaviour* 39 (1990): 960–66.

Boyles, J. G., P. M. Cryan, G. F. McCracken, and T. H. Kunz. "Economic Importance of Bats in Agriculture." *Science*, 332 (2011): 41–42.

Boyles, J. G., C. L. Sole, P. M. Cryan, and G. F. McCracken. "On Estimating the Economic Value of Insectivorous Bats: Prospects and Priorities for Biologists." In *Bat Evolution, Ecology and Conservation*, edited by R. A. Adams and S. C. Pedersen. New York and London: Springer, 2013.

Cleveland, C. J., M. Betke, P. Federico, J. D. Frank, T. G. Hallam, J. Horn, J. D. Lopez, Jr., G. F. McCracken, R. A. Medellin, A. Moreno-Valdez, C. G. Sansone, J. K. Westbrook, and T. H. Kunz. "Economic Value of Pest Control Service Provided by Brazilian Free-Tailed Bats in South-Central Texas." *Frontiers in Ecology and the Environment* 4, no. 5 (2006): 238–43.

Davis, R. B., C. F. Herreid II, and H. L. Short. "Mexican Free-Tailed Bats in Texas." *Ecological Monographs* 32 (1962): 311–46.

Des Marais, D. J., J. M. Mitchell, W. G. Meinschein, and J. M. Hayes. "The Carbon Isotope Biochemistry of the Individual Hydrocarbons in Bat Guano and the Ecology of the Insectivorous Bats in the Region of Carlsbad, New Mexico." *Geochimica et Cosmochimica Acta* 44 (1980): 2075–86.

Gustin, M. K., and G. F. McCracken. "Scent Recognition Between Females and Pups in the Bat *Tadarida brasiliensis Mexicana*." *Animal Behaviour* 35 (1987): 13–19.

Kalka, M. B., A. R. Smith, and E.K.V. Kalko. "Bats Limit Arthropods and Herbivory in a Tropical Forest." *Science* 320 (2008): 71.

Keeley, B. W., and M. D. Tuttle. *Bats in American Bridges*. Austin, TX: Bat Conservation International, Resource Publication no. 4, 1999.

Leelapaibul, W., S. Bumrungsri, and A. Pattanawiboon. "Diet of Wrinkle-Lipped Free-Tailed Bat (*Tadarida plicata* Buchannan, 1800) in Central Thailand: Insectivorous Bats Potentially Act as Biological Pest Control Agents." *Acta Chiropterologica* 7, no. 1 (2005): 111–19.

Loughry, W. J., and G. F. McCracken. "Factors Influencing Female-Pup Recognition in Mexican Free-Tailed Bats." *Journal of Mammalogy* 72, no. 3 (1991): 624–26.

McCracken, G. F. "Communal Nursing in Mexican Free-Tailed Bat Maternity Colonies." *Science* 223 (1984): 1090–91.

McCracken, G. F., and J. K. Westbrook. "Bat Patrol." *National Geographic* 201, no. 4 (2002): 116–23.

Mizutani, H., D. A. McFarlane, and Y. Kabaya. "Nitrogen and Carbon Isotope Study of Bat Guano Core from Eagle Creek Cave, Arizona, U.S.A." *Mass Spectroscopy* 40 (1992): 57–65.

Taylor, P. J., K. Bohmann, J. N. Steyn, M. C. Schoeman, E. Matamba, M. Zepeda-Mendoza, T. Nangammbi, and M.T.P. Gilbert. "Bats Eat Pest Green Vegetable Stink Bugs (*Nezara viridula*): Diet Analyses of Seven Insectivorous Species of Bats Roosting and Foraging in Macadamia Orchards." In *Southern African Macadamia Growers' Association Yearbook*. Tzaneen, South Africa: SAMAC, 2013.

Taylor, P. J., A. Monadjem, and J. N. Steyn. "Seasonal Patterns of Habitat Use by Insectivorous Bats in a Subtropical African Agro-Ecosystem Dominated by Macadamia Orchards." *African Journal of Ecology* 51 (2013): 552–61.

Tuttle, M. D. "The Lives of Mexican Free-Tailed Bats." *BATS* 12, no. 3 (1994): 6–14.

Williams-Guillen, K., I. Perfecto, and J. Vandermeer. "Bats Limit Insects in a Neotropical Agroforestry System." *Science* 320 (2008): 70.

CHAPTER 10: AFRICAN ADVENTURES

Fenton, M. B., and J. L. Eger. "*Chaerephon chapini*." *Mammalian Species*, no. 692: 1–2.

Ryan, M. J., and M. D. Tuttle. "The Role of Prey-Generated Sounds, Vision and Echolocation in Prey Localization by the African Bat *Cardioderma cor* (megadermatidae)." *Journal of Comparative Physiology* 161 (1987): 59–66.

Srinivas, G. "*Lavia frons* (Yellow-Winged Bat)." Animal Diversity Web, University of Michigan Museum of Zoology, 2002. http://animaldiversity.ummz.umich.edu/accounts/Lavia_frons/.

Vaughan, T. "Nocturnal Behavior of the African False Vampire Bat (*Cardioderma cor*)." *Journal of Mammalogy* 57, no. 2 (1976): 227–48.

Yates, C. "In Loving Memory of Our Friend Paul Kabochi." Monkey Trek, 2003. http://www.monkeytrek.com/news/africa_first/paul.htm.

CHAPTER 11: BAT-LOVING MONKS, TIGERS, AND POACHERS

Ballenger, L. "*Kerivoula picta* (Painted Bat)." Animal Diversity Web, University of Michigan Museum of Zoology, 1999. http://animaldiversity.org/site/accounts/information/Kerivoula_picta.html.

Bumrungsri, S., E. Sripaoraya, T. Chongsiri, K. Sridith, and P. Racey. "The Pollination Ecology of Durian (*Durio zibethinus, Bombacaceae*) in Southern Thailand." *Journal of Tropical Ecology* 25 (2009): 85–92.

Burns, A. "*Craseonycteris thonglongyai* (Bumblebee Bat)." Animal Diversity Web, University of Michigan Museum of Zoology, 2013. http://animaldiversity.ummz.umich.edu/accounts/Craseonycteris_thonglongyai/.

Frances, C. M. *A Guide to the Mammals of Southeast Asia.* Princeton, NJ: Princeton University Press, 2008.

Leelapaibul, W., S. Bumrungsri, and A. Pattanawiboon. "Diet of Wrinkle-Lipped Free-Tailed Bat (*Tadarida plicata* Buchannan, 1800) in Central Thailand: Insectivorous Bats Potentially Act As Biological Pest Control Agents." *Acta Chiropterologica* 7, no. 1 (2005): 111–19.

Tuttle, M. D. "Harmless, Highly Beneficial, Bats Still Get a Bum Rap." *Smithsonian* 14, no. 10 (1984): 74–81.

Wanger, T. C., K. Darras, S. Bumrungsri, and T. Tscharntke. "Bat pest control contributes to food security in Thailand." *Biological Conservation* 171 (2014): 220–23.

CHAPTER 12: MYSTERIES OF BAT-GUIDING FLOWERS

McGrath, S. "Call of the Bloom." *National Geographic* 225, no. 3 (2014): 128–39.

Simon, R., M. W. Holderied, C. U. Koch, and O. von Helversen. "Floral Acoustics: Conspicuous Echoes of a Dish-Shaped Leaf Attract Bat Pollinators." *Science* 333, no. 6042 (2011): 631–33.

Von Helversen, D., M. W. Hoderied, and O. von Helversen. "Echoes of Bat-Pollinated Bell-Shaped Flowers Conspicuous for Nectar-Feeding Bats?" *Journal of Experimental Biology* 206 (2003): 1025–34.

Von Helversen, D., and O. von Helversen. "Acoustic Guide in Bat-Pollinated Flower." *Nature* 398 (1999): 759–60.

CHAPTER 13: BAT FORESTERS

Bumrungsri, S., A. Harbit, C. Benzie, K. Carmouche, K. Sridith, and P. Racey. "The Pollination Ecology of Two Species of *Parkia* (Mimosaceae) in Southern Thailand." *Journal of Tropical Ecology* 24 (2008): 467–75.

Bumrungsri, S., E. Sripaoraya, T. Chongsiri, K. Sridith, and P. Racey. "The Pollination Ecology of Durian (*Durio zibethinus, Bombacaceae*) in Southern Thailand." *Journal of Tropical Ecology* 25 (2009): 85–92.

Fleming, T. H. "Fruit Bats: Prime Movers of Tropical Seeds. *BATS* 5, no. 3 (1987): 3–5.

Fujita, M. S. "Flying Foxes and Economics." *BATS* 6, no. 1 (1988): 4–9.
Fujita, M. S., and M. D. Tuttle. "Flying Foxes (*Chiroptera Pteropodidae*): Threatened Animals of Key Ecological and Economic Importance." *Conservation Biology* 5 (1991): 1–9.
Graham, G. L., and M. Murphy. "The Changing Face of Bat Conservation in the Pacific." *BATS* 8, no. 1 (1990): 3–5.
Hodgkison, R., S. T. Balding, A. Zubaid, and T. H. Kunz. "Fruit Bats (*Chiroptera: Pteropodidae*) as Seed Dispersers and Pollinators in a Lowland Malaysian Rain Forest." *Biotropica* 35, no. 4 (2003): 491–532.
Lobova, T. A., C. K. Geiselman, and S. A. Mori. *Seed Dispersal by Bats in the Neotropics*. Bronx, NY: New York Botanical Garden, 2009.
McConkey, K. R., and D. R. Drake. "Flying Foxes Cease to Function as Seed Dispersers Long Before They Become Rare." *Ecology* 87 (2006): 271–76.
Muscarella, R., and T. H. Fleming. "The Role of Frugivorous Bats in Tropical Forest Succession." *Biological Review* 82 (2007): 1–18.
Pettigrew, J. D. "Flying Primates? Megabats Have the Advanced Pathway from Eye to Midbrain." *Science* 231, no. 4743 (1986): 1304–06.
Racey, P. "Eight Million Fruit Bats: Africa's Best Kept Secret." *BATS* 22, no. 1 (2004): 1–5.
Ratcliffe, F. N. *The Flying Fox* (Pteropus) *in Australia*. Commonwealth of Australia, Council for Scientific and Industrial Research, Bulletin no. 53 (1931): 1–81.
Srithongchuay, T., S. Bumrungsri, and E. Sripao-raya. "The Pollination Ecology of the Late-Successional Tree, *Oroxylum indicum (Bignoniaceae)* in Thailand." *Journal of Tropical Ecology* 24 (2008): 477–84.
Thomas, D. W. "On Fruits, Seeds and Bats." *BATS* 9, no. 4 (1991): 8–13.
Tuttle, M. D. "Gentle Fliers of the African Night." *National Geographic* 169, no. 4 (1986): 540–58.

CHAPTER 14: FRUIT GROWER COMPLAINTS

Cardozo, G. "Flying Fox Researcher Says Plans to Give Councils More Power to cull bats Won't Work." *Daily Telegraph*, November 7, 2014. http://www.dailytelegraph.com.au/newslocal/central

-coast/flying-fox-researcher-says-plans-to-give-councils-more-power-to-cull-bats-wont-work/story-fngr8h0p-1227114177186?nk.

Condon, M. "New South Wales Bat Cull Could Start This Week, But Vets Warn It Won't Work." ABC Rural, October 26, 2014. http://www.abc.net.au/news/2014-10-27/nrn-nsw-bat-cull-warning-27-10-14/5843570.

Makin, D., and H. Mendelssohn. "Insectivorous Bats Victims of Israeli Campaign." *BATS* 2, no. 4 (1985): 1–2 and 3.

Ratcliffe, F. N. *The Flying Fox* (Pteropus) *in Australia.* Commonwealth of Australia, Council for Scientific and Industrial Research, Bulletin no. 53 (1931): 1–81.

Tuttle, M. D. "Fruit Bats Exonerated." *BATS* 1, no. 2 (June 1984). http://www.batcon.org/resources/media-education/bats-magazine/bat_article/193.

CHAPTER 15: A NATIONAL PARK FOR BATS

Holmes, S. "*Pteropus samoensis* (Samoan Flying Fox)." Animal Diversity Web, University of Michigan Museum of Zoology, 2002. http://animaldiversity.org/accounts/Pteropus_samoensis/.

National Park Service. "National Park of American Samoa." http://www.nps.gov/npsa/index.htm.

Stein, C. "A National Park for Bats." *BATS* 15, no. 4 (1997): 12–13.

EPILOGUE: HOPE FOR THE FUTURE

Arnett, E. B., and E. F. Baerwald. "Impacts of Wind Energy Development on Bats: Implications for Conservation." In *Bat Evolution, Ecology and Conservation,* edited by R. A. Adams and S. C. Pedersen. New York and London: Springer, 2013, pp. 435–56.

Kunz, T. H., E. B. Arnett, W. P. Erickson, A. R. Hoar, G. D. Johnson, R. P. Larkin, M. D. Strickland, R. W. Thresher, and M. D. Tuttle. "Ecological Impacts of Wind Energy Development on Bats: Questions, Research Needs, and Hypotheses." *Frontiers in Ecology and the Environment* 5, no. 6 (2007), pp. 315–24.

Reeder, D. M., and M. S. Moore. "White-Nose Syndrome: A Deadly Emerging Infectious Disease of Hibernating Bats." In *Bat Evolution, Ecology and Conservation,* edited by R. A. Adams and S. C. Pedersen. New York and London: Springer, 2013, pp. 413–34.

Tuttle, M. D. "Threats to Bats and Educational Challenges." In *Bat Evolution, Ecology and Conservation*, edited by R. A. Adams and S. C. Pedersen. New York and London: Springer, 2013, pp. 363–91.

——. "Saving America's Beleaguered Bats." *National Geographic* 188, no. 2 (1995): 36–57.

INDEX